AF307210

Progress in Molecular and Subcellular Biology

Series Editors: W.E.G. Müller (Managing Editor), Ph. Jeanteur, I. Kostovic, Y. Kuchino, A. Macieira-Coelho, R.E. Rhoads

25

Springer

Berlin
Heidelberg
New York
Barcelona
Hong Kong
London
Milan
Paris
Singapore
Tokyo

Alvaro Macieira-Coelho (Ed.)

Signaling Through the Cell Matrix

With 22 Figures

Springer

Professor Dr. Alvaro Macieira-Coelho
University of Paris VI
73 bis rue Maréchal Foch
78000 Versailles
France

ISSN 0079-6484
ISBN-13: 978-3-642-64117-6 Springer-Verlag Berlin Heidelberg New York

Library of Congress Cataloging-in-Publication Data

Signaling through the cell matrix / Alvaro Macieira-Coelho (ed.). –
Berlin; Heidelberg; New York; Barcelona; Hong Kong; London;
Milan; Paris; Singapore; Tokyo: Springer 2000
  (Progress in molecular and subcellular biology; 25)
  ISBN-13: 978-3-642-64117-6   e-ISBN-13: 978-3-642-59766-4
  DOI: 10.1007/978-3-642-59766-4

Springer-Verlag is a company in the BertelsmannSpringer publishing group.
© Springer-Verlag Berlin Heidelberg 2000
Softcover reprint of the hardcover 1st edition 2000

Cover design: Meta Design, Berlin
Typesetting: Best-set Typesetter Ltd., Hong Kong
SPIN 10730885 39/3130 – 5 4 3 2 1 0 – Printed on acid-free paper

Preface

The genetic information contained in a cell needs the appropriate environment to express itself, not only the intracellular environment but also the extracellular one. The latter is provided to a great extent by the molecules which constitute the extracellular matrix.

On the one hand, the matrix creates inter alia the right pH and osmotic environment and allows the diffusion of messengers targeting the cell membrane; on the other hand, it has a mechanical effect whose relevance began to be understood 28 years ago.

Basically, the messages that reach the cell and are then transported to the genome depend on molecular conformational flexibility. Molecular structures usually prevail because they represent states of minimum potential energy creating energy barriers which are activated through conformational changes. From the periphery to the nucleus the information flows through the activation of energy barriers. The tools used to switch from low-energy to high-energy molecular configurations are: the binding of ligands to their receptors, gradients of electrochemical potential created by ion pumps, Ca2+ mobilization, and phosphorylation and dephosphorylation. Variation in molecular configuration through molecular binding is in itself sufficient to trigger ion pumps and activate kinases and phosphatases. This is one aspect of the mechanical role of the extracellular matrix dealt with herein: the induction of molecular and supramolecular conformational modifications through interactions with the cell membrane, which promote the transduction and centripetal progression of signals.

This is not necessarily accompanied by changes in cell shape. Cell shape is just the visible part of the iceberg. The fundamental control of the transmission of the information depends on the way in which the intracellular network is connected, i.e., its topology. New topological constraints must be created which can occur without modifications of cell shape. The experiments that demonstrated this feature are presented in this Volume. They concern a very basic regulation of cell function, the transfer of energy; much of the future of cell biology depends on understanding this processing of information.

Moreover, membranes are structures far from equilibrium. Instants of stabilization have to be created to order substrates in advantageous configurations and to anchor cell membrane regions, thus increasing the probability of messengers finding their targets on the membrane. This is achieved through the interaction of molecules from the extracellular matrix with other molecules in the cell membrane such as integrins. This is another important aspect focused in this Volume.

Furthermore, the Volume illustrates the relationship of causality between these events brought about by the interaction of the cell with its matrix, and the activation of gene expression. Through the creation of molecular configurations favorable for molecular binding, genes are transcribed or repressed, and proteins are synthesized and degraded, sending the message back centrifugally.

Finally, the description of different cell types, normal and malignant, gives the reader a broad picture of the role of these crucial regulations in the physiology and pathology of the organism.

Versailles, April 2000   ALVARO MACIEIRA-COELHO

Contents

Role of Focal Adhesion Kinase in Signaling by the Extracellular Matrix
Ji-He Zhao and Jun-Lin Guan

Interaction Between Cells and Extracellular Matrix: Signaling by Integrins and the Elastin-Laminin Receptor
J. Labat-Robert and L. Robert

Regulation of Gene Expression by Changes in Cell Adhesion
T. Kuzumaki

**Expression of Liver Specific-Genes in Hepatocytes
Cultured in Collagen Gel Matrix**
M.J. Gomez-Lechon, R. Jover, T. Donato, X. Ponsoda, and J.V. Castell

Collagen Type I: A Substrate and a Signal for Invasion
L. Van Hoorde, E. Van Aken, and M. Mareel

Topological Constraints Carry Signaling from the Cell Matrix to the Genome

Alvaro Macieira-Coelho[1]

1
Variation in Cell Adhesion During Proliferation of Normal Cells

Cell division is a source of cellular drift. Indeed, a daughter cell is not the exact copy of the mother cell and is not exactly identical to the sister cell (Macieira-Coelho et al. 1982). The small genomic modifications created by cell division are cumulative and become apparent morphologically, structurally, and metabolically (Macieira-Coelho 1988a).

This phenomenon has been particularly well studied in normal somatic mammalian cells proliferating in vitro. Genomic changes become progressively manifest: DNA is not distributed symmetrically in a significant fraction of the cell population during cell division (Macieira-Coelho et al. 1982), it is progressively reorganized at the different hierarchical orders of its structure (Macieira-Coelho 1988b), and there are changes in gene expression (Pignolo et al. 1993).

The genomic modifications have repercussions on cell structure and morphology. The organization of the cytoskeleton is modified, molecular changes occur in the cell membrane, and the interaction of the cell with its substratum evolves. Opinions diverge whether this evolution of somatic cell populations through proliferation is a programmed (Martin et al. 1974) or a stochastic degenerative (Rubin 1997) phenomenon. Whatever its nature, it constitutes an excellent tool to understand the phenotypic evolution of a cell population through division.

Cell attachment depends on the electric charges at the cell surface, and, indeed, measurements of the electrophoretic mobility revealed that the net negative surface charge decreases during proliferation in vitro (Bosmann et al. 1976). This is in agreement with the decrease of membrane-bound sialic acid (Milo and Hart 1976), the molecule that is mainly responsible for the negative charge expressed at the cell surface.

Several attempts have been made to find a biochemical basis for the changes in plasma membrane attachment to the substratum. They were first made on chicken fibroblasts by Courtois and Hughes (1974), who measured the

---

[1] University of Paris VI, 73 bis rue Maréchal Foch, 78000 Versailles, France

Progress in Molecular and Subcellular Biology, Vol. 25
A. Macieira-Coelho (Ed.)

incorporation of radioactive D-glucosamine into cellular macromolecules. Labeled substances removed by trypsin from the cell surface were analyzed, as well as the smooth membrane fractions isolated from trypsinized cells. The surface membrane of cells with a longer proliferative history was associated with larger amounts of proteoglycans and lesser amounts of several glycopeptides.

Fry and Weisman-Shomer (1977) and Aizawa et al. (1980) found a decrease in a 220 000-MW cell surface protein in human fibroblasts after serial proliferation.

Schachtschabel and Wever (1978), Wever et al. (1980), and Sluke et al. (1981) examined the synthesis and distribution pattern of glycosaminoglycans. A reduced rate of incorporation of radioactive sulfate and glucosamine occurred at the end of the proliferative cell life span, with a relatively greater decline of [^{35}S]sulfate incorporation than of [^{14}C]glucosamine into extracellular glycosaminoglycans. The authors suggested that both the decline of glycosaminoglycan synthesis and the sharp fall in the growth potential are associated phenomena; it gives further support to the concept of the existence of a terminal phase (phase IV) (Macieira-Coelho and Taboury 1982) in which there is a profound disorganization of proliferation.

The relationship between growth decline and surface proteoglycans was also investigated by Matuoka and Mitsui (1981). They confirmed that the relative amount of heparane sulfate increases in cells at the end of their proliferative life span, and that this change is inversely correlated with the saturation density. Confluent glutaraldehyde-fixed, late-passage cells were inhibitory for the growth of early-passage cells, and treatments that degraded heparan sulfate removed the inhibitory activity. The authors suggested that heparan sulfate is involved in the increased sensitivity to contact inhbition of growth known to occur during serial proliferation (Macieira-Coelho et al. 1966a).

Vogel and coworkers (1981) concluded that secreted and cell-surface glycosaminoglycans represent different pools and that in late-passage cells there is an effect primarily on the secreted pool. Since glycosaminoglycans play a role in the binding of fibronectin to collagen, the alterations described could be involved in the absence of fibronectin at the surface of cells at the end of their proliferative life span (Vogel et al. 1981). On the other hand, increases in the overall synthesis of collagen and in the ratio of the alpha-1 and alpha-2 chains of type I collagen have been reported in rat fibroblasts (Konterman and Bayreuther 1979), changes that could contribute to the diminished binding of fibronectin and render attachment looser. Late-passage cells take a longer time to spread on the substratum, occupy a larger area, and have more attachment sites but a weaker binding force. The increased area could be a compensation for the lower binding force.

Chandrasekhar et al. (1983) found that during serial proliferation fibroblasts show a decreased cell-substratum adhesion. The main defect did not seem to depend on the membrane itself, but rather upon the secretion of molecules responsible for attachment. Indeed, cells could recover the adhesivity toward the substratum when incubated with medium conditioned by cells with a shorter proliferative lifetime or with fibronectin purified from this same medium. Membrane alterations, however, must take place, since cells that had lost adhesivity needed more fibronectin for attachment. Qualitative differences were revealed on SDS-polyacrylamide gels, in proteins precipitated from medium conditioned by cells with a longer proliferative history, suggesting that fibronectin becomes defective. Little or no fibronectin was found on the surfaces of substratum-attached cells (Vogel et al. 1981) and the biosynthesis of fibronectin increased as a function of passage number (Rasoamanantena et al. 1994). There are multiple causes for the alteration of this critical molecule involved in cell attachment, inter alia alterations in the elastin-laminin receptor which controls the biosynthetic mechanisms of some constituents of the extracellular matrix (Fodil-Bourhala et al. 1999).

Cell adhesion depends also on the integrity of the cytoskeleton and of the molecules establishing the link between the cytoskeleton and external molecules. Integrins play a crucial role in this link. The polarized topography of integrins is gradually lost through proliferation (Luca et al. 1992). Moreover, levels of the alpha-5-beta-1 fibronectin receptor, a member of the integrin family, are reduced at the end of the cell population proliferative life span (Hu et al. 1996). Results also suggested a sequestration of the alpha-5 polypeptide subunit.

Alterations of the cytoskeleton were first reported by Bowman and Daniel (1975), who observed with scanning electron microscopy that postmitotic cells lacked prominent bundles of microfilaments and that this deficiency coincided with reduced motility and decreased ability to incorporate 3H-thymidine.

Anderson (1978) measured the amount of actin and analyzed its peptide composition in chicken fibroblasts. He found an increased actin content in cells that had lost the division potential, mainly of one protein species homologous to muscle actin.

Kelley et al. (1980) studied the relationship between cell adhesion and the cytoskeleton and concluded that cell spreading is prolonged in late-passage cells and is correlated with a retarded assembly of actin bundles. In addition, detergent-resistant membrane regions that are associated with underlying bundles of filaments take a longer time to develop. The actin fibers (Kelley et al. 1980) and the microtubules are differently organized, with a loss of polarity in the latter (Raes et al. 1984; Van Gansen et al. 1984). This is expressed as a centrifugal depolymerization of the microtubules, contrary to cells that still retain their proliferative potential, where depolymerizaton is centripetal.

One can infer that the modifications in cell-substratum adhesion through proliferation are most likely caused by a disturbance in the interaction of surface molecules with membrane molecules and the cytoskeleton, whose synthesis is differently coordinated. They are strictly coupled with modifications of the kinetics of cell proliferation.

1.1
Coupling Between Changes in Cell Adhesion
and Proliferation of Normal Cells

The modifications of cell adhesion are accompanied by an increase in cell volume (Simons 1967). This was detected measuring cell diameters with the aid of an ocular micrometer and later confirmed with electronic counters (Macieira-Coelho and Pontén 1969; Simons and Van den Broeck 1970; Mitsui and Schneider 1976; Whatley and Hill 1979).

In chicken fibroblasts, increases in cell volume are coupled with decline in the probability of cycling (Lima and Macieira-Coelho 1972). The first detectable increase in volume took place around the 15th cell population doubling and coincided with a prolongation of the time needed to reach the maximal DNA synthetic activity, which was reached 12h after seeding the cells instead of 6h, followed by a prolongation of the doubling time. Later, towards the end of proliferative life span, cell size increased again just before the post-mitotic state was reached.

The relationship between changes in cell volume and cell division was further approached using different methodologies. Tritiated thymidine labeling (Bowman et al. 1975), direct analysis by cinematography (Absher and Absher 1976), and comparison of cell volume with population doubling time (Mitsui and Schneider 1976) all led to the conclusion that the increase in cell volume coincides with a decreased probability of entering the division cycle. Higher growth factor concentrations and larger substratum areas become necessary for proliferation (Collins et al. 1979).

The decline of the net negative surface charge observed during the proliferative life span of human fibroblasts follows the pattern of the decline of the rate of entrance into DNA synthesis after seeding the cells on a substratum (Macieira-Coelho and Azzarone 1982). Changes in surface charge increase the time needed to reach the right amount of spreading necessary to initiate DNA synthesis. These events are related with specific alterations in the display of surface oligosacharides which mediate the initial adherence of the cells to the substratum (Mann et al. 1992).

When cells reach the end of their proliferative life span, the enlargement becomes very profound, with striking changes in cell morphology (Fig. 1). The cells become flat with few microvilli, show little ruffling activity, and are almost completely devoid of macropinocytosis (Blomquist et al. 1978). The cytoplasm and the nuclear and nucleolar areas increase simultaneously (BeMiller and Miller 1979). This is accompanied by dramatic changes in the organization of DNA (Macieira-Coelho 1995) and of gene expression (Hara et al. 1993).

Under electron microscopy the analysis of the organization of the 30-nm chromatin fibers revealed that the density of chromatin threads declines progressively (Macieira-Coelho 1991). At the very end, the entrance into a post-mitotic state coincides with a dramatic fall in the spacing between the fibers, especially at the nuclear periphery, where they are shorter and unusually

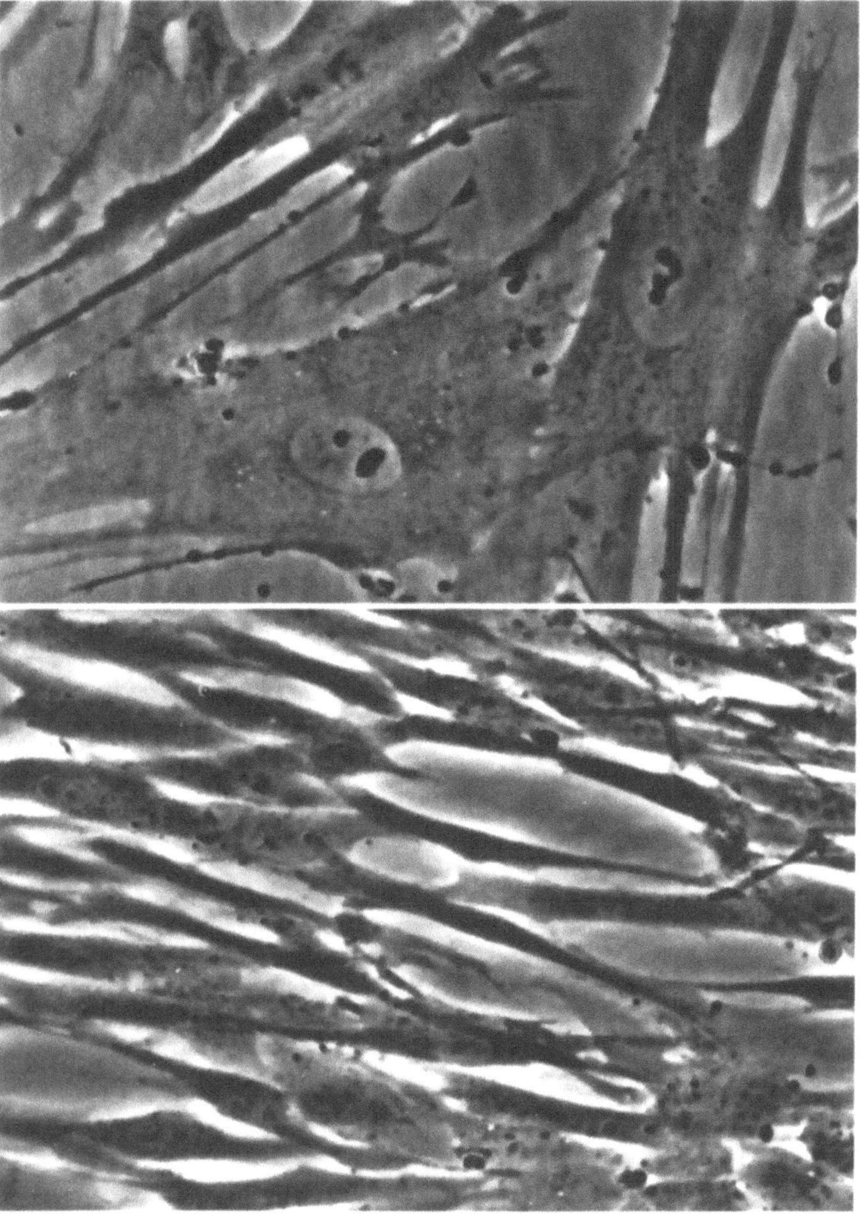

Fig. 1. Human fibroblasts in the proliferative phase (*below*) and in the postmitotic terminal phase (*above*)

spaced along the lamina densa, which sometimes is entirely devoid of chromatin threads.

The pattern of evolution of the density and spacing of the chromatin fibers has analogies with that of the fraction of cells synthesizing DNA during a 24-h period (Macieira-Coelho 1991). The percentage of cells synthesizing DNA during the first 24h after seeding them on a plastic surface declines progressively through the cell population life span, like the density of the chromatin fibers. The spacing, however, increases significantly only during the last divisions, in an inverse relation with the maximal percentage of cells synthesizing DNA during a 24-h period between seeding the cells and the time when they reach resting phase (Macieira-Coelho and Azzarone 1982). These data show that as the cells progress through their life span it takes longer to build up the adhesion sites, to attach and to spread, so that the initiation of the division cycle is increasingly delayed after having seeded the cells. This suggests that the delay in the organization of the cytoskeleton for cell attachment is associated with changes in the anchorage of chromatin and with the initiation of DNA synthesis.

The cell structural modifications are accompanied by a decline in the ability to contract evaluated by the capacity to retract a plasma clot (Macieira-Coelho and Azzarone 1990). The decline in cell contractility occurred in parallel with the decline in the maximal number of cells capable of synthesizing DNA during a 24-h period between seeding the cells and the time they reached resting phase (Macieira-Coelho and Azzarone 1982).

So the data show that decreased adhesion, decreased contractility, and increase in cell size are due to a different organization of the cellular scaffold, which is accompanied by a reorganization of the chromatin fibers coupled with an increasing difficulty in initiating DNA synthesis. This cycle of events can be summarized as follows:

Genome reorganization during cell division

↓

Reorganization of the cellular scaffold

↓

Changes in cell attachment, spreading, and DNA anchorage

↓

Decreased flexibility of the cytoplasm and nuclear cage

↓

Reorganization in the high-order structure of DNA

↓

Decreased probability of initiating the division cycle

Cell movements are necessary for the initiation of the division cycle and of DNA synthesis because of the topological changes they create at the molecular and supramolecular levels. Indeed, membrane movements transmitted through the cytoskeleton to the nuclear cage help to give to DNA the right conformation for gene expression during the G1 period, and to initiating sites the right steric configuration for the synthesis of DNA (Macieira-Coelho 1983). With the discoordination of the synthesis of the components of the cellular scaffold, the cell increases in size and progressively loses its contractility, so that the flow of movements from the periphery to the nucleus declines. Cell size in itself must create profound changes through modification of the flexibility of conformation, and of the association between molecules. The changes in surface and intracellular molecules will feed back on the genome and accentuate its reorganization.

2
Coupling Between Changes in Cell Adhesion and Proliferation of Transformed Cells

Viral-transformed and tumor-derived cell populations cultivated in vitro display a loss of contact inhibition of growth that can be measured relating the percentage of dividing cells with the cell density (Macieira-Coelho 1967a,b). The phenomenon, however, can be expressed differently because of changes in cell adhesion during transformation (Fig. 2).

In human fibroblasts transformed by Rous sarcoma virus (RSV), the growth curve reaches a plateau and the percent of cells synthesizing DNA falls to low levels, but at a cell density twice that of the controls (Macieira-Coelho 1967a). In RSV-transformed bovine fibroblasts, the cells reach a high density with a growth curve that never reaches a plateau, and the percentage of cells synthesizing DNA never falls below 50%. In SV40-transformed human fibroblasts the maximum cell density is not above that of the controls cultures, the growth curve assumes a sigmoid shape, and the percentage of cells synthesizing DNA never falls below 50%.

So these transformed cultures have in common the fact that when they reach the density where the respective controls enter resting phase, the percentage of dividing cells is still significantly high (Macieira-Coelho 1967b); the transformed phenotype, however, is very different. This can be explained in terms of the same phenomenon, loss of contact inhibition of growth, expressed differently because of different affinities of the cells towards neighboring cells and towards the substratum (Fig. 2). Human fibroblasts transformed with RSV have very little tendency to overlap neighboring cells but have a strong affinity towards the solid substratum (Macieira-Coelho 1967a). Thus, although they lose contact inhibition of growth, DNA synthesis stops when the cells are immobilized because each cell is attached to the minimum area possible for survival after occupying all the surface of the culture vessel. Bovine fibroblasts infected with RSV adhere well to the solid substratum and have a great tendency to overlap neighboring cells (Stenkvist 1966), so that DNA synthesis does not stop, because the cells growing on top of one another have always new substratum available to proliferate. Human fibroblasts transformed by SV40 do not have a tendency to overlap; since adhesion to the substratum is weak, cells

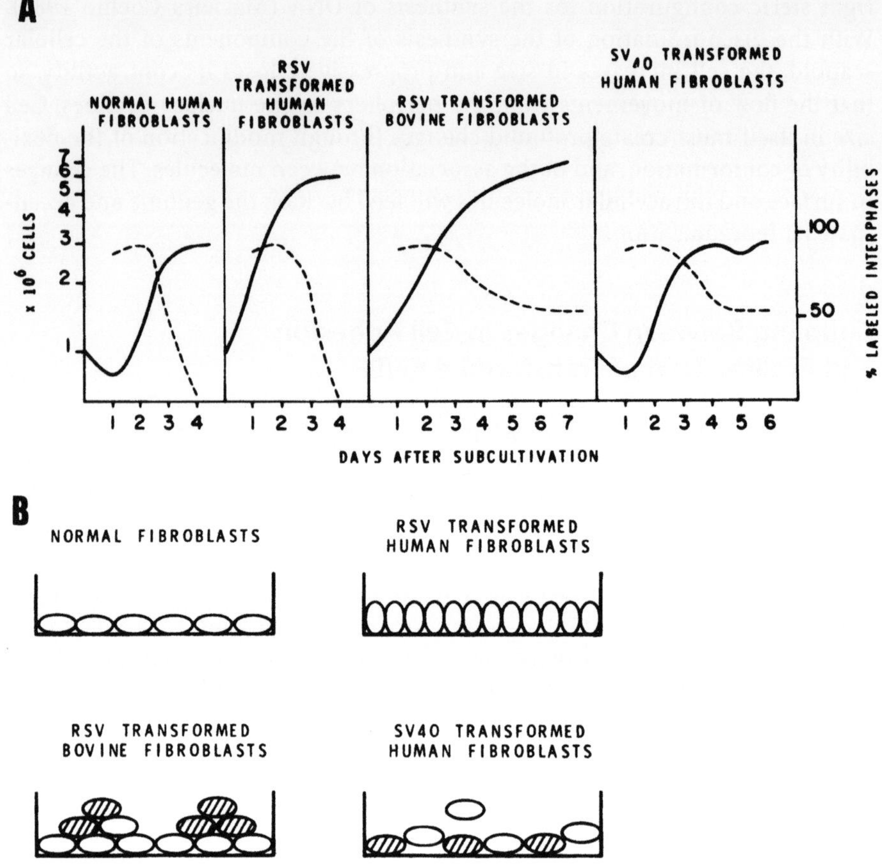

Fig. 2. A Growth (——) and DNA synthesis (----) curves of normal and virus transformed fibroblasts. **B** Schematic representation of the behavior of the same cells growing on a plastic surface. For further explanation see text

detach continuously, leaving new areas available for attachment and continuous proliferation.

In conclusion, the genetic modifications induced by the viruses modify the synthesis of molecules, leading to structural modifications and new constraints in cell attachment that contribute to the deregulation of growth. The respective transformed phenotype, however, depends upon how cell adhesion is modified.

3
Mechanisms of Cell Adhesion-Mediated Changes in the Cell Phenotype

The data described above show that new properties develop in normal and transformed cells through variations in cell structure and adhesion. These occur through genomic modifications caused by serial divisions in normal cells and by the virus during transformation. The results suggest that it should be possible to modulate the cell phenoype via a substratum where the physicochemical properties regulating cell attachment could be controlled. The following experimental system was devised to demonstrate this assumption.

A protein polymer made of bovine serum albumin (BSA) was adapted to cell culture using a method available for the preparation of insoluble protein polymers (Macieira-Coelho and Avrameas 1972). The polymerization of BSA is due to cross-linkage originated by bonds between amino groups of proteins and active aldehyde groups of glutaraldehyde. After polymerization of the protein, active aldehyde groups from glutaraldehyde remain free in the polymer. Substances possessing free amino groups (e.g. polylysine) will link by covalent bonds to these aldehyde goups. Substances like heparin or DEAE-dextran which do not possess free amino groups are presumably fixed to the BSA polymer by noncovalent bonds; heparin binds to positively charged groups (e.g. amino groups) and DEAE-dextran to negatively charged groups (e.g. carboxyl groups) of the BSA. In this way the protein polymer can be covered with a variety of substances that give different physicochemical properties to the surface used to support cell growth (Macieira-Coelho and Avrameas 1973).

3.1
Modulation of Cell Proliferation

The transformed mouse L cell line, which on conventional cell culture flasks grows as a monolayer, was seeded on neutral or negatively charged polymers obtained by covering the BSA with polytryptophane, polyleucine, polyphenylalanine, polytyrosine, polycysteine, or polyglutamic acid. On all these surfaces the cells proliferated in clusters, reaching a higher number than when growing as a monolayer (Macieira-Coelho and Avrameas 1972). Positively charged surfaces were obtained covering the BSA with polylysine, polyornithine, polyhistidine, polyarginine, or DEAE-dextran. On these surfaces the L cells formed monolayers and proliferated to higher densities than on plastic. Figure 3 illustrates the morphology of sparse cells on these surfaces. Introducing negatively charged groups (e.g., heparin) on positively charged surfaces, the cells still formed monolayers but the final cell yield was decreased (Macieira-Coelho and Avrameas 1972).

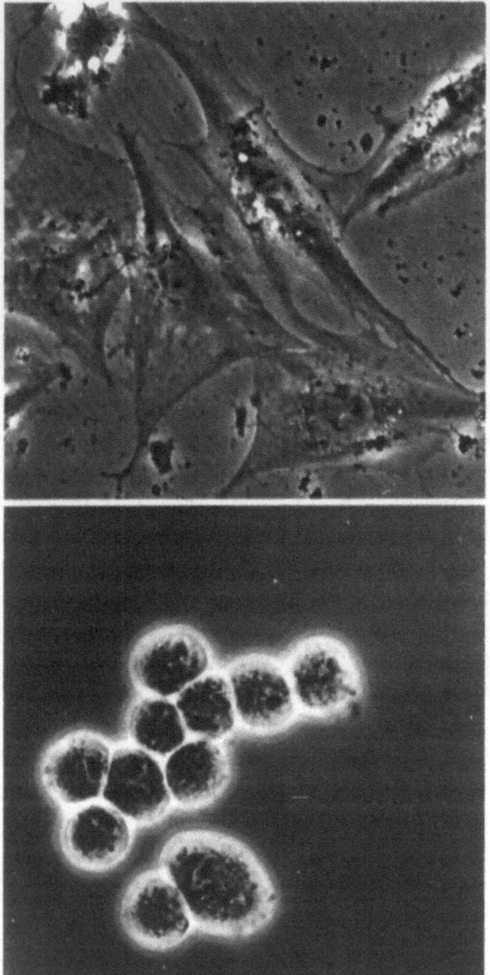

Fig. 3. Morphology of mouse L cells seeded on a negatively (*below*) and a positively (*above*) charged surface

The hamster BHK cell line also formed clusters on negatively charged and monolayers on positively charged surfaces, but proliferated very little when in clusters (Macieira-Coelho et al. 1974).

Carter (1965) showed that cells in culture tend to migrate to the substratum with which they have a better affinity. This property, called haptotaxis, can explain the aggregation or formation of monolayers on different surfaces. L and BHK cells on the negatively charged surfaces have a greater affinity towards each other than towards the solid substratum, and thus will form clusters instead of monolayers. Growth under these conditions, however, will depend on the capacity of the cells to proliferate using another cell as substratum which is the case for L cells but not for BHK. These data and those described in the previous section show that contact inhibition of growth is substratum-dependent.

Monolayering or aggregation is due to the electric charge of the substratum, since all neutral or negatively charged substances caused rounding of the cells, while all positively charged ones induced spreading, independently of other properties of those substances (Macieira-Coelho et al. 1974). Furthermore, neither the degree of binding of the polyamino acid to the polymer nor structural differences influenced the cell behavior. The critical nature of the electric charge could also be demonstrated binding to the BSA, polyamino acids composed of different proportions of lysine and alanine. The number of attached cells had a direct relationship with the amount of lysine (Macieira-Coelho et al. 1974).

The polymer had no direct effect on cell metabolism. Indeed, substances bound to the substratum did not enter the cell. Moreover, substances that are able to influence cell metabolism when in solution, such as polyornithine, which stimulates the transport of proteins, lose this property when in an insoluble form bound to the substratum (Macieira-Coelho et al. 1974). Results also showed that polymerized serum used as a substratum loses its nutritional and growth-promoting properties on cultivated cells (Macieira-Coelho et al. 1974), and that growth factors lose their mitogenic property when attached to the substratum (Avrameas et al. 1976).

Finally, both l and d forms of the polyamino acids could equally modulate the cell phenotypes described above, again suggesting that the substratum had no direct effect on cell metabolism (Macieira-Coelho et al. 1974).

3.2
Modulation of Malignancy

L-1210 mouse leukemia cells which normally grow in suspension were seeded on a BSA polymer positively charged (Macieira-Coelho and Avrameas 1972). A week later, the cells progressively lost their rounded shape and spread out on the polymer, assuming a fibroblastic form. The fibroblastic cells progressively invaded the whole surface while the round cells were lost during subcultivation. The cells that acquired adhesive properties had lost their malignant potential (Macieira-Coelho and Avrameas 1973).

The fibroblastic L-1210 cells were serially passaged during 1 year. After this time, round cells growing in suspension could again be recovered and carried separately (Macieira-Coelho 1988b). This population had reacquired its malignant properties, although not fully, since 10^6 cells took longer to kill the animals, and one survived. The same amount of cells from the parent leukemic cell line rapidly killed all the animals. These results suggest that different genes are expressed when cells either attach and spread or grow in suspension.

3.3
Modulation of Cell Differentiation

The effect of these substrata on cell differentiation was tested using myogenic cells of the rat L6 cell line which proliferate and fuse in culture to form myotubes that actively synthesize myosine.

The cells plated on negatively charged BSA (covered with polyglutamic acid) started growing with a doubling time of 20 to 22h and formed a confluent monolayer on the 3rd to 4th day after plating (Wahrmann et al. 1981). Fusion began on the 4th to 5th day and continued until the 8th day, when 60 to 80% of all nuclei were included in myotubes. Myosin synthesis increased abruptly at the beginning of the formation of myotubes and progressed thereafter.

The cells plated on positively charged BSA (covered with poylysine) proliferated with a doubling time of 48 to 50h and formed a confluent monolayer on the 4th to 5th day. Multinucleated myotubes, however, could not be observed even after 8 to 10 days in culture. The relative rate of myosin synthesis in cells grown on this surface started increasing earlier than in cultures grown on the negatively charged surface, but it reached lower values. The inhibition of cell differentiation was coupled with a restraint of DNA synthesis and cloning efficiency (Wahrmann et al. 1981).

Creatine kinase was determined in the myoblast cultures plated on polyglutamic acid and polylysine-coated surfaces. The activity of the enzyme increased in parallel to cell fusion on the former surface; in cells plated on positively charged surfaces, the enzyme activity began earlier but reached final lower values (Sénéchal et al. 1984).

The measurement of cloning efficiency of the myogenic cells grown on negatively and positively charged surfaces revealed that on the former the cloning efficency fell to 10% on the 6th day after plating the cells, when differentiation was completed. An identical fraction of L6 cells is unable to differentiate. On the polylysine-coated surface, on the 6th day after subcultivation, 30–50% of the cells still had the potential to divide, judged by their cloning efficiency, although growth was arrested, as revealed by the DNA synthesis curve and the amount of DNA per culture (Wahrmann et al. 1981).

Scanning electron microscopy revealed striking differences between the myoblast seeded on a negatively or a positively charged surface. They concerned the cell surface, which was smooth on the former substratum, contrary to that of cells on polylysine, where the surface was covered with blebs and villosities (Fig. 4).

The effect of reducing the net positive charge of the cell attachment surface was evaluated by adding a neutral amino acid (tryptophan) before treating the BSA polymer with polylysine. The morphology of the cells was identical to that of those grown on a surface covered with polylysine alone. Differentiation progressed further than on polylysine alone but then aborted, establishing a direct correlation between the presence of polylysine and the aborted differentiation (Wahrmann et al. 1981).

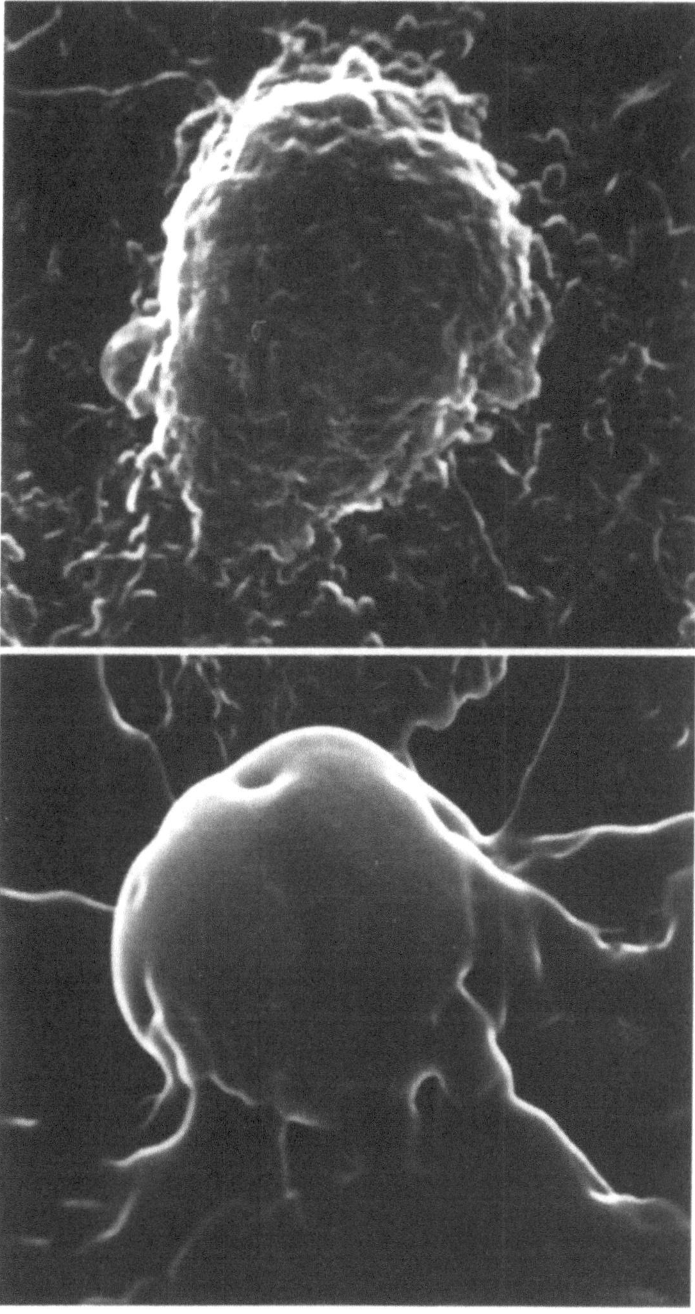

Fig. 4. Myoblast seeded on a negatively (*below*) and a positively (*above*) charged polymer

It could be postulated that the observed effects were not due to the electrical charge of the substratum but rather that the latter binds selectively divalent cations, thus changing the ionic composition of the medium. This can be excluded because in the absence of cells the conductivity of the medium, which was 10.5 microsiemens, remained unchanged even after a 4-h incubation at 37 °C on either of the surfaces used (Wahrmann et al. 1981).

It could also be argued that the cells do not interact with the charged surfaces themselves but with serum proteins selectively adsorbed to them. To test this hypothesis noncoated as well as polylysine- and polyglutamic acid-coated BSA polymers were incubated with culture medium during 4h at 37 °C before exposure to the cells. Subsequently the medium was discarded, the substrata were extensively washed with PBS and incubated further with a solution of 10% sodium lauryl sulfate during 10 min at 37 °C. No released proteins could be detected in the supernatants that had been in contact with these surfaces. A small quantity of proteins was solubilized and separated by SDS-polyacrylamide gel electrophoresis. Only one polypeptide migrating identically to native BSA could be identified (Wahrmann et al. 1981).

To determine whether the arrest of differentiation is due to the effect of the charged surface or to changes in cell shape, the myoblasts were plated and grown on plastic surfaces. Polylysine-covered beads were added each day after subcultivation to half of the cultures (Sénéchal et al. 1984). In the cultures without beads differentiation proceeded with myosin synthesis and the formation of myotubes. In cultures to which polylysine-coated beads were added, the beads attached firmly to the cells and could be detached neither by vigorous shaking nor by washing the cultures. The morphology of the cells was the same as in the control cultures growing on plastic, without the spreading characteristics of cells plated on polylysine-coated BSA. Nevertheless addition of the beads aborted the differentiation process. Polyglutamic acid-coated beads did not stop differentiation (Sénéchal et al. 1984).

These results show that spreading per se or changes in cell shape is not the limiting factor for myoblast division and differentiation; the physical action of the electric charge is the effector.

4
Conclusions

Serially proliferating somatic cells go through genomic reorganizations which induce a progressive structural evolution changing cell volume and adhesion to a substratum. The modifications of the probability of cycling are strictly coupled with the structural evolution and the adhesion to the substratum. On the other hand, the genomic modifications introduced by oncogenic viruses create different adhesive properties that correlate with the transformed phenotype.

The experimental system described above where the cell phenotype could be modulated seeding the cells on substrata with different physicochemical properties, illustrates how adhesion can send the information to the genome. Indeed, the modulation of differentiation and, in particular, malignancy for several cell generations can only be explained by an effect on gene expression of the cell adhesion-dependent signaling.

It could be postulated that it is the modification of cell morphology that conditions the cell phenotype. The experiments show, however, that proliferation and differentiation can be modulated without modifications of cell shape, and that physical forces constitute the signal triggered at the membrane that modifies the cell genotype and phenotype. This can be explained by the structural organization of the animal cell.

Mammalian DNA is a molecule more than 1 m long, confined in a sphere, the nucleus, with a diameter of about 5 microns. This implies an elaborate folding, which is regulated inter alia by the anchorage of this makeup to a protein matrix. The scaffold upon which chromatin is anchored plays a crucial role in the organization of the high-order structure of DNA. There is indeed a protein framework, called the nuclear protein matrix, with which DNA is associated. DNA synthesis-initiating sites are preferentially located at the borders between condensed chromatin and interchromatin areas (Berezney and Coffey 1975), suggesting an important role for the nuclear matrix, in particular the peripheral nuclear region, in the initiation of DNA replication. The anchorage of DNA is crucial not only for replication but also for transcription, since nascent RNA is associated with the nuclear cage (Jackson et al. 1984).

The nuclear lamina, a filamentous protein meshwork lining the nucleoplasmic surface of the nuclear envelope, probably provides an anchoring site at the nuclear periphery for interphase chromatin (Gerace 1985). When the nuclear shell is isolated, it contains chromatin structures made of packed nucleosomes 28–32-nm thick (identical to the high-order solenoid DNA structure) that are associated with the three nuclear lamins (Bouvier et al. 1985). The lamina is composed of proteins called lamins, which seem to be intermediary structures between DNA-binding proteins and the cytoskeleton. Indeed, on the one hand, the lamina is tightly bound to chromatin since it can be dissociated from chromatin only by high salt solution which also extracts the tightly bound histones in the nucleosome cores (Bouvier et al. 1985). On the other hand, lamins have a striking sequence homology with intermediate filaments, a component of the cytoskeleton (Gerace 1985).

Thus, the anchorage of chromatin seems to be fulfilled with the preservation of the continuity with the cytoplasmic scaffold. In this way, DNA is linked to the cytoskeleton through its anchorage to the nuclear cage and via the former to the cell membrane and the extracellular matrix. The cytoskeleton and the nuclear matrix act as integrators not only of space but also of function within the cell. This whole structure can be seen as a three-dimensional manifold (Thurston and Weeks 1984) constituted by linked cranks, where information flows to a great extent through changes in molecular configurations which are not necessarily accompanied by changes in cell shape.

Variation in molecular configuration through physical forces is in itself sufficient to trigger aggregation of molecules, ion pumps, and phosphorylation cascades. On the other hand, that is also what the latter do, switching molecules from low- to high-energy configurations, thus activating energy barriers.

Through the creation of molecular configurations favorable for molecular binding, genes are transcribed or repressed. In substratum-dependent animal cells, adhesion is the trigger to build up tension within the cell network, whose function is regulated through the synthesis of molecules with the right steric configuration and through energy turnover. Cell behavior is determined by the way in which this network is connected, i.e., by its topology. Changes in cell adhesion modulate gene expression and cell phenotype via new topological constraints.

References

Absher PM, Absher RG (1976) Clonal variation and aging of diploid fibroblasts. Cinematographic studies of cell pedigrees. Exp Cell Res 103:247–255

Aizawa S, Mitsui Y, Kurimoto F, Matsuoko K (1980) Cell-surface changes accompanying aging in diploid fibroblasts. Effect of tissue, donor age and genotype. Mech Ageing Dev 13:297–306

Anderson PJ (1978) Actin in young and senescent fibroblasts. Biochem J 160:169–172

Avrameas S, Ternynck T, Macieira-Coelho A (1976) Loss of mitogenic activity by immobilized lectins. Biochem Biophys Res Commun 72:790–795

BeMiller PM, Miller JE (1979) Cytological changes in senescing WI-38 cells. A statistical analysis. Mech Ageing Dev 10:1–15

Berezney R, Coffey DS (1975) Nuclear protein matrix: association with newly synthesized DNA. Science 189:291–293

Blomqvist E, Arro E, Brunk U, Westermark B (1978) Plasma membrane motility of cultured human glia cells in phase II and III. Acta Pathol Microbiol Scand A Pathol 86:257–263

Bosmann HB, Gutheil RL Jr, Case KR (1976) Loss of a critical neutral protease in ageing WI-38 cells. Nature 261:499–501

Bouvier D, Hubert J, Seve AP, Bouteille M (1985) Characterization of lamina-bound chromatin in the nuclear shell isolated from HeLa cells. Exp Cell Res 156:500–512

Bowman PD, Daniel CW (1975) Aging of human fibroblasts in vitro, surface features and behaviour of aging WI-38 cells. Mech Ageing Dev 4:147–158

Bowman PD, Meek RL, Daniel CW (1975) Aging of human fibroblasts in vitro, correlation between DNA synthetic ability and cell size. Exp Cell Res 93:184–190

Carter SB (1965) Principles of cell mobility: the direction of cell movement and cancer invasion. Nature 208:1183–1187

Chandrasekhar S, Norton E, Millis AJT, Izzard CS (1983) Functional changes in cellular fibronectin from late-passage fibroblasts in vitro. Cell Biol Int Rep 7:11–21

Collins VP, Arro E, Blomqvist E, Brunk U, Frederikson BA, Westermark B (1979) Cell locomotion and proliferation in relation to available surface area, serum concentration and culture age. Scanning Electron Microsc 111:411–420

Courtois Y, Hughes RC (1974) Glycoproteins of chick embryo fibroblasts in cultures with a finite life span. Eur J Biochem 44:131–138

Courtois Y, Weisman-Shomer P (1977) Surface proteins of young and senescent cultured avian fibroblasts. Cell Biol Int Rep 1:399–407

Fodil-Bourhala I, Drubaix I, Robert L, Labat-Robert J (1999) Effect of in vitro aging on the modulation of protein and fibronectin biosynthesis by the elastin-laminin receptor in human skin fibroblasts. Gerontology 45:23–30

Fry M, Weisman-Shomer P (1977) Surface proteins of young and senescent cultured avian fibroblasts. Cell Biol Int Rep 1:399-407

Gerace L (1985) Structural proteins in the eukaryotic nucleus. Nature 318:508-509

Hara E, Yamaguchi T, Tahara H, Tsuyama N, Tsurui H, Ide T, Oda K (1993) DNA-DNA subtractive cDNA cloning using oligo(dT)-latex and PCR: identification of cellular genes which are overexpressed in senescent human diploid fibroblasts. Anal Biochem 214:58-63

Hu Q, Moerman EJ, Goldstein S (1996) Altered expression and regulation of the alpha-5-beta-1 integrin-fibronectin receptor lead to reduced amounts of functional alpha-5-beta-1 heterodimer on the plasma membrane of senescent human diploid fibroblasts. Exp Cell Res 224:251-263

Jackson DA, McCready SJ, Cook PR (1984) Replication and transcription depend on attachment of DNA to the nuclear cage. J Cell Sci Suppl 1:59-79

Kelley RO, Trotter JA, Marek LF, Perdue BD, Taylor CB (1980) Variation in cytoskeletal assembly during spreading of progressively subcultivated human fibroblasts (IMR-90). Mech Ageing Dev 13:127-141

Konterman K, Bayreuther K (1979) The cellular aging of rat fibroblasts in vitro is a differentiation process. Gerontology 25:261-270

Lima L, Macieira-Coelho A (1972) Parameters of aging in chicken embryo fibroblasts cultivated in vitro. Exp Cell Res 70:279-284

Luca M de, Pellegrini G, Bondanza S, Cremona O, Savoia P, Cancedda R, Marchisio PC (1992) The control of polarized integrin topography and the organization of adhesion-related cytoskeleton in normal human keratinocytes depend upon number of passages in culture and ionic environment. Exp Cell Res 202:142-150

Macieira-Coelho A (1967a) Dissociation between inhibition of movement and inhibition of division in RSV transformed human fibroblasts. Exp Cell Res 47:193-200

Macieira-Coelho A (1967b) Relationship between DNA synthesis and cell density in normal and virus transformed cells. Int J Cancer 2:296-303

Macieira-Coelho A (1983) Changes in membrane properties associated with cellular aging. Int Rev Cytol 83:183-220

Macieira-Coelho A (1988a) Biology of normal proliferating cells in vitro. Relevance for in vivo aging. In: von Hahn HP (ed) Interdisciplinary Topics in Gerontology, vol 23. Karger, Basel, pp 1-172

Macieira-Coelho A (1988b) Intracellular signal flow and genome activation. In: Bergener M, Ermini M, Stähelin HB (eds) Crossroads in aging. Academic Press, London, pp 3-23

Macieira-Coelho A (1990a) Reorganization in the different hierarchical structures of DNA during cell senescence. In: Finch CE, Johnson CE (eds) Molecular biology of aging, vol 123. Wiley-Liss, New York, pp 351-364

Macieira-Coelho A (1990b) Cancer and aging at the cellular level. In: Macieira-Coelho A, Nordenskjöld B (eds) Cancer and aging. CRC Press, Boca Raton, pp 11-37

Macieira-Coelho A (1991) Chromatin reorganization during senescence of proliferating cells. Mutat Res 256:81

Macieira-Coelho A (1995) The last mitoses of the human fibroblast proliferative life span, physiopathological implications. Mech Ageing Dev 82:91-104

Macieira-Coelho A, Avrameas S (1972) Modulation of cell behavior in vitro by the substratum in fibroblastic and leukemic mouse cell lines. Proc Natl Acad Sc USA 69:2469-2473

Macieira-Coelho A, Avrameas S (1973) Protein polymers as a substratum for the modulation of cell poliferation in vitro. In: Kruse PF, Patterson MK (eds) Tissue culture, methods and applications. Academic Press, New York, pp 379-383

Macieira-Coelho A, Azzarone B (1982) Aging of human fibroblasts is a succession of subtle changes in the cell cycle and has a final short stage with abrupt events. Exp Cell Res 141:325-332

Macieira-Coelho A, Azzarone B (1990) Correlation between contractility and proliferation in human fibroblasts. J Cell Phys 142:610-614

Macieira-Coelho A, Pontén J (1969) Analogy in growth between late passage human embryonic and early passage human adult fibroblasts. J Cell Biol 43:374–377

Macieira-Coelho A, Taboury F (1982) A revaluation of the changes in proliferation in human fibroblasts during ageing in vitro. Cell Tissue Kinet 15:213–224

Macieira-Coelho A, Pontén J, Philipson L (1966a) The division cycle and RNA synthesis in diploid human cells at diffeent passage levels in vitro. Exp Cell Res 43:20–29

Macieira-Coelho A, Pontén J, Philipson L (1966b) Inhibition of the division cycle in confluent cultures of human fibroblasts in vitro. Exp Cell Res 43:673–684

Macieira-Coelho A, Berumen L, Avrameas S (1974) Properties of protein polymers as substratum for cell growth in vitro. J Cell Phys 83:379–388

Macieira-Coelho A, Bengtson A, van der Ploeg M (1982) Distribution of DNA between sister cells during serial subcultivation of human fibroblasts. Histochemistry 75:11–24

Mann PL, Busse SC, Griffey RH, Tellez CM (1992) Cell surface oligosacharide modulation during differentiation: V. Partial characterization of the regulated surface during substrate adhesion and spreading. Mech Ageing Dev 62:47–77

Martin GM, Sprague CA, Norwood TH, Pendergrass WR (1974) Clonal selection, attenuation and differentiation in an in vitro model of hyperplasia. Am J Pathol 74:137

Matuoka K, Mitsui Y (1981) Involvement of cell surface heparan sulfate in density dependent inhibition of cell proliferation. Cell Struct Funct 6:23–34

Milo GE, Hart RW (1976) Age-related alterations in plasma membrane glycoprotein content and scheduled or nonscheduled DNA syntheis. Arch Biochem Biophys 176:324–333

Mitsui Y, Schneider EL (1976) Relationship between cell replication and volume in senescent human diploid fibroblasts. Mech Ageing Dev 5:45–56

Pignolo RJ, Cristofalo VJ, Rotenberg MO (1993) Senescent WI-38 cells fail to express EPC-1, a gene induced in young cells upon entry into the Go state. J Biol Chem 268:89–49

Raes M, de Brabander M, Remacle J (1984) Polyploid cells in ageing hamster fibroblasts in vitro: possible implication of the centrosome. Eur J Cell Biol 35:79–80

Rasoamanantena R, Thweatt R, Labat-Robert J, Goldstein S (1994) Altered regulation of fibronectin gene expression in Werner syndrome fibroblasts. Exp Cell Res 213:121–127

Rubin H (1997) Cell aging in vivo and in vitro. Mech Ageing Dev 98:1–35

Schachtschabel DO, Wever J (1978) Age-related decline in the synthesis of glycosaminoglycans by cultured human fibroblasts (WI-38). Mech Ageing Dev 8:257–264

Sénéchal H, Wahrmann JP, Delain D, Macieira-Coelho A (1984) Modulation of differentiation in vitro. II. Influence of cell spreading and surface events on myogenesis. In Vitro 20: 692–698

Simons JW (1967) The use of frequency distributions of cell diameters to characterize cell populations in tissue culture. Exp Cell Res 45:336–350

Simons JW, Van den Broek C (1970) Comparison of ageing in vitro and and ageing in vivo by means of cell size analysis using a Coulter counter. Gerontologia 16:340–351

Sluke G, Schachtschabel DO, Wever J (1981) Age-related changes in the distribution pattern of glycosaminoglycans synthesized by cultured human diploid fibroblasts (WI-38). Mech Ageing Dev 16:19–27

Stenkvist B (1966) Long-term cultivation of human and bovine fibroblastic cells morphologically transformed in vitro by Rous sarcoma virus. Acta Pathol Microbiol Scand 67:67–82

Thurston PW, Weeks JR (1984) The mathematics of three-dimensional manifolds. Sci Am July:94–106

Van Gansen P, Siebertz B, Capone B, Malherbe L (1984) Relationship between cytoplasmic microtubular complex, DNA synthesis and cell morphology in mouse embryonic fibroblasts (effects of age, serum deprivation, aphidicolin, cyochalasin B and colchicine). Biol Cell 52:161–174

Vogel KG, Kelley RO, Steward C (1981) Loss of organized fibronectin matrix from the surface of aging diploid fibroblasts. Mech Ageing Dev 16:295–303

Wahrmann JP, Delain D, Bournoutian C, Macieira-Coelho A (1981) Modulation of differentiation in vitro. I Influence of the attachment surface on myogenesis. In Vitro 17:752–762

Wever J, Schachtschabel DO, Sluke G, Wever G (1980) Effect of short- or long-term treatment with exogenous glycosaminoglycans on growth and glycosaminoglycan synthesis of human fibroblasts (WI-38) in culture. Mech Ageing Dev 14:89–99
Whatley SA, Hill BT (1979) The relationship between DNA content, cell volume and growth potential in ageing human embryonic mesenchymal cells. Cell Biol Int Rep 3:671–683

The Transmission of Contractility Through Cell Adhesion

Kyoko Imanaka-Yoshida[1]

1
Introduction

Living cells generate active tension within their internal cytoskeleton and exert on the adhesive sites with extra cellular matrix (ECM). This mechanical linkage is necessary not only for supporting the cell shape or position, but also for cell migration. (Harris, Stopak et al. 1981; Sheetz 1994; Lauffenburger and Horwitz 1996; Mitchison and Cramer 1996) Furthermore, it is becoming clear that the internal stress field of the cell, resulting from the balancing of generation and exertion forces, can influence a number of cellular functions, including growth, differentiation, apoptosis, and signal tranduction (reviewed in Shyy and Chien 1997; Chicurel et al. 1997). Cells are also interconnected with neighboring cells and transmitting forces to their neighbors. These mechanical interactions between groups of cells and the ECM, which joins these cells together, are important in tissue structure maintenance, physiological organ function, morphogenesis, and wound healing.

This chapter will focus on the mechanical interaction between contractility generated by cells and ECM transmitted through cell adhesion sites.

2
Generation of Contractility

2.1
Stress Fibers

The force-generating mechanism of cytoskeletal contractility is considered to be an actin-myosin interaction of stress fibers (Harris 1994; Burridge and Chrzanowska-Wodnicka 1996). Stress fibers contain many of the contractile proteins, such as actin and myosin II, which are arranged in a sarcomeric configuration. As this structural pattern suggests, stress fibers have contractile

---

[1] Department of Pathology, Mie University School of Medicine, 2-174 Edobashi, Tsu, Mie 514-8507, Japan

Progress in Molecular and Subcellular Biology, Vol. 25
A. Macieira-Coelho (Ed.)
© Springer-Verlag Berlin Heidelberg 2000

potential. Because stress fibers are usually linked to a rigid substratum, they generate isometric tension termed traction force. Traction forces have been experimentally visualized and measured (Harris, Wild et al. 1980; Danowski and Harris 1988; Danowski 1989; Lee et al. 1994; Oliver et al. 1995). For example, when cells that have prominent stress fibers are grown on flexible silicone rubber surfaces, they visibly wrinkle these substrates. Detachment, or loss of stress fibers in cells, results in the loss of wrinkles. The traction forces deform collagen gels, reflecting rearrangement of collagen fibrils in vivo. (Stopak and Harris 1982; Stopak et al. 1985).

2.2
Formation of Stress Fibers and Focal Adhesion

Application of tension to the cell in culture induces a bundling of actin filaments and stimulates stress fibers. (Franke et al. 1984; Kolega 1986)). When fibroblasts are cultured in three-dimensional free-floating collagen gels, the cells contract the gel. If the gels are anchored to inhibit contraction, the fibroblasts develop prominent stress fibers and generate isometric tension (Mochitate et al. 1991; Tomasek et al. 1992; Grinnell 1994; Halliday and Tomasek 1995). Following release of tension in anchored gels, there is a rapid contraction of gel, followed by the disappearance of stress fibers (Tomasek and Akiyama 1992; Grinnell 1994). Therefore, isometric tension generated by stress fibers contributes to the formation of stress fibers themselves, which requires a bidirectional transmission of the contractility from inside the cell to connective tissue and external tension to the cell.

3
Molecular Mechanism of Force Transmission

3.1
Focal Adhesions

Both inward and outward mechanical force transmission should occur at physically linked adhesion sites. One model system for studying cell adhesion to the ECM is the focal adhesion. Focal adhesions are specialized sites of adhesion developed by many cells in culture. They consist of transmembranous ECM receptors (integrin) interacting with exterior ECM components and interior actin filaments. Many proteins, including α-actinin, vinculin, talin, focal adhesion kinase (FAK), paxillin, and tensin, have been identified at focal adhesions (Jockusch et al. 1995). Focal adhesions are thought to be sites of physical linkage between actin filaments and integrin, as well as initiation sites of various signal transductions (reviewed in Burridge and Chrzanowska-Wodnicka 1996).

3.2
Mechanism of Force Transmission

One model to explain the force transmission mechanism is tensegrity. The tensegrity model suggests that the cell is rigidly connected, or hard-wired to nuclei and to the ECM by a cytoskeletal tensegrity structure, so that cells can immediately respond to mechanical stress. (Ingber 1997; Wang et al. 1993; Ingber et al. 1994). In this model, the cell is presumed to coordinate actin-myosin filament generating tension and microtubules, which resist compression and stabilize against buckling by tension bearing intermediate filaments. When external mechanical force is transmitted through a surface receptor (integrin), which physically couples the actin cytoskeleton to the ECM, it is transduced throughout inside the cells via a cytoskeletal network. The finding that the pulling of integrin molecules by micromanipulation causes reorientation of cytoskeletons, distortion of nuclei, and redistribution of nucleoli along the axis-applied tension supports this hypothesis (Maniotis et al. 1997).

Another possibility is that mechanical stress can be translated to biochemical signals. Several reports suggest the existence of a signaling molecule, which mediates the interaction between tension and microtubules. (Bershadsky et al. 1996; Danowski 1998). Furthermore, it is well established that the focal adhesion complex may represent a major site for signal integration between growth factors and the ECM-based signaling pathway. Mechanical force transmission via focal adhesion may stimulate the signaling molecules localized at this site and cause responses common to those activated by growth factors or ECM bindings. For example, mechanical stretch increases phosphorylation of the focal adhesion tyrosine kinase, pp125FAK (Hamasaki et al. 1995)

3.3
Molecules Mediating Force Transmission

In either model, it is clear that molecules forming the adhesion complex could mediate bidirectional transmission of external stress to the cell and force generated inside the cell to outside. As discussed in the tensegrity model, $\beta1$ integrins can act as mechanoreceptors and transmit mechanical signals to cytoskeletons (Maniotis et al. 1997). Conversely, several reports have demonstrated that integrins mediate the transmission of mechanical tension generated within the cytoskeleton to the ECM. For example, anti-$\beta1$ integrin antibody inhibits collagen gel contraction by fibroblasts stimulated with PDGF or angiotesin II (Gullberg et al. 1990; Burgess et al. 1994). Various integrins, including integrin $\alpha1\beta1$ (Racine-Samson et al. 1997), $\alpha2\beta1$ (Klein et al. 1991; Schiro et al. 1991; Riikonen et al. 1995), and $\alpha6\beta1$ (Imanaka-Yoshida et al. 1999), possibly dependent on cell type, can mediate the transmission of cytoskeletal contractility to connective tissue.

Beyond the ECM-receptor, other cytoskeletal components of focal adhesions, such as vinculin, may also participate in the force transmission (Danowski et al. 1992; Ezzell et al. 1997; Goldmann et al. 1998; Imanaka-Yoshida et al. 1999). The absence of vinculin correlates with the decrease in the mechanical stiffness of integrin-cytoskeletal linkage. (Ezzell et al. 1997; Goldmann et al. 1998)

4
Regulation of Contractility Transmission

The organization of the actin cytoskeleton and the assembly of focal adhesions should be major determinants of the contractile forces exerted against the ECM.

4.1
Regulation of Contractile Force

4.1.1
The Regulatory Mechanism of RhoA

Although multiple regulatory mechanisms have been studied, one of the major regulators of focal adhesions and stress fibers is the GTP-binding protein, RhoA. Activated RhoA induces the assembly of stress fibers and focal adhesions (see review in Burridge and Chrzanowska-Wodnicka 1996; Schoenwaelder and Burridge 1999). RhoA.GTP binds and activates several serine/threonine kinases, such as Rho-kinase, ROCKII and ROKα, phosphorylates and inhibits myosin phosphatase, and results in elevated myosin light chain phosphorylation (Kimura et al. 1996). Myosin light chain phosphorylation, in turn, promotes both myosin filament assembly and actin-activated myosin ATPase activity, resulting in the bundling of actin filaments. Tension is subsequently generated and transmitted to integrins, which promotes aggregation of diffusely distributed integrin into focal adhesions. (Chrzanowska-Wodnicka and Burridge 1996)

4.1.2
Involvement of Microtubules

It is suggested that microtubules also play a role in regulating contractility and the organization of actin cytoskeletons and focal adhesion assembly. For instance, microtubules depolymerization causes rapid formation of stress fibers and adhesion plaques in phorbol ester-treated fibroblasts and increase in the contractility of fibroblasts grown on flexible silicone rubber substrata or collagen gel. (Danowski 1989)

The stimulation of contractility upon depolymerization of microtubules may be explained by the tensegrity model (Ingber 1993; Ingber et al. 1994), however, there is also a possibility that disruption of microtubules initiates a signaling release of microtubule-associated proteins (Danowski 1998). Recent studies highlighted the possibility that microtubule disruption either directly or indirectly activates the RhoA signaling pathway, because disruption of microtubules leads to phosphorylation of the myosin light chain (Kolodney and Elson 1995), Rho activation (Enomoto 1996; Liu et al. 1998), and phosphorylation of paxillin and FAK (Bershadsky et al. 1996). These two molecules are also phosphorylated upon RhoA activation (Chrzanowska-Wodnicka and Burridge 1994; Seufferlein and Rozengurt 1994; Flinn and Ridley 1996), as well as in response to cell-ECM adhesion by integrin (Burridge et al. 1992). It seems likely that microtubule depolymerization could activate a signaling pathway similar to integrin-mediated signal transduction.

4.2
Regulation of Stability of Focal Adhesions

Both the assembly and disassembly of focal adhesion, but also disassembly of focal adhesion, should affect force transmission. Besides inducing focal adhesion formation, strong tension generated by cells also causes the release of adhesion (Crowley and Horwitz 1995). Furthermore, other regulatory mechanisms mediated by tyrosine phosphorylation, protein kinase C (Schliwa et al. 1984), or protein kinase A (Lamb et al. 1988; Lampugnani et al. 1990; Glass and Kreisberg 1993) are involved in the assembly/disassembly of focal adhesions.

Linkage strength among molecules comprising the adhesion complex would be another determinant of contractile transmission. Binding of vinculin to talin and actin is regulated by phosphoinositides (Johnson and Craig 1994) and inhibited by acidic phospholipid. (Weekes et al. 1996).

5
Interaction Between Contraction Force and ECM

The ECM is another regulator of contractile force transmission. For example, several ECM proteins such as thromspondin or tenascin-C can promote disassembly of focal adhesions and stress fibers (Murphy-Ullrich and Höök 1989; Murphy-Ullrich et al. 1991; 1996). Cellular fibronectin is necessary for contraction of collagen gel by fibroblasts (Asaga et al. 1991). Interestingly, a recent study has shown that matrix rigidity and components regulate the strength of integrin-cytoskeleton linkages (Choquet et al. 1997), which suggests that cells regulate contractility depending on the extracellular environment.

It is important to note that cells synthesize, degrade, and reorganize extracellular matrix components. This remodeling process is controlled by many factors, including mechanical stress. Several reports have demonstrated that mechanical forces regulate the gene expression of ECM proteins ((Hatamochi et al. 1989; Carver et al. 1991; Lambert et al. 1992; Eckes et al. 1993; Chiquet-Ehrismann et al. 1994; Trächslin et al. 1999).

Physical forces not only control the transcription of genes but they can also directly affect ECM molecules to control their assembly and organization. The molecular mechanisms implicated in the control by forces of the assembly of ECM proteins have been studied mainly with fibronectin. Recently, Zhang et al. (1997) suggested that fibronectin matrix assembly depends on RhoA-mediated contractility. Tension generated by the cytoskeleton contributes to the assembly of a fibrillar matrix fibronectin (Halliday and Tomasek 1995; Zhong et al. 1998). Disruption of microfilaments with cytochalasin disrupts fibrillar fibronectin. Fibronectin matrix assembly is promoted by LPA, a stimulator of RhoA activity in serum (Zhang et al. 1994). Furthermore, using GFP-tagged fibronectin, Ohashi et al. observed that cells exert tension and stretching of fibronectin (Ohashi et al. 1999). These findings suggest the possibility that tension generated by cytoskeleton and mediated by RhoA can stretch the fibronectin molecule to expose the cryptic assembly sites, and in turn, enhance fibronectin fibril formation.

6
Myofibroblasts and Wound Healing

Most of the cellular and molecular mechanisms of generation and transmission of contractility have been based on cultured cells. One of the most prominent *in vivo* examples of the transmission of contractility are myofibroblasts in wound healing (reviewed in Grinnell 1994). During wound healing, fibroblasts differentiate into myofibroblasts, which express high levels of α-smooth muscle actin (Darby et al. 1990; Arora and McCulloch 1994), and generate tissue contraction. The myofibroblasts organize a well-developed actin cytoskeleton along the line of greatest tension (Petroll et al. 1993) and adhere tightly to the surrounding fibronectin-rich granulation tissue (Singer et al. 1984; Tomasek et al. 1992). These structures are quite similar to the stress fibers and focal adhesion observed in cultured cells.

Using floating versus anchored collagen gel culture systems as models of wound contraction, many studies have shown that various external factors, such as TGF-β, PDGF, FGF, interferon γ and endothelin, positively or negatively control the contractility (reviewed in Grinnell 1994) mediated by several signaling pathways (Grinnell et al. 1999). Furthermore, it is also suggested that tension generated by myofibroblasts may control proliferation, apoptosis, differentiation, and contractility of myofibroblasts in granulation tissue (Arora et al. 1999; Grinnell et al. 1999).

7
Muscle

Skeletal and cardiac muscle cells are highly differentiated for contractility and have well-developed cytoskeletal system and subsarcolemmal cytoskeletons for contraction and contractile transmission (reviewed in Small et al. 1992; Berthier and Blaineau 1997).

7.1
Myotendinous Junctions

Myotendinous junctions are major force transmission sites in skeletal myocytes to connective tissue (tendon) (Tidball and Daniel 1986). The edge of myofibrils inside the myocytes are linked through the cell membrane to the ECM at myotendinous junctions, which are structurally very similar to focal adhesions and stress fibers. Many proteins at focal adhesions, including cytoskeletal proteins such as vinculin, talin (Tidball and Daniel 1986; Frenette and Tidball 1998), and integrin (Bozyczko et al. 1989; Swasdison and Mayne 1989), are concentrated at this junction and are thought to anchor myofibrils to the ECM via interactions with laminin, heparan sulfate proteoglycan, or tenascin-C (Bozyczko et al. 1989; Swasdison and Mayne 1989; Engvall et al. 1990). A signaling molecule of focal adhesion sites, focal adhesion kinase, is also concentrated as focal adhesions (Baker et al. 1994). Dystrophin is also present in the myotendinous junction (Byers et al. 1991). In the field of muscle research, particular interest has been focused on dystrophin, since this molecule is absent or altered in Duchenne muscle dystrophy (Hoffman et al. 1987; Campbell 1995). Dystrophin is associated with subsarcolemmal actin filaments, dystrophin-associated glycoproteins, and the ECM; it is thought to provide stability to the sarcolemma by protecting it from mechanical stresses during contraction-relaxation cycles (Matsumura and Campbell 1994; Tinsley et al. 1994).

7.2
Costameres

7.2.1
Contraction Force Transmission at Costameres

Costameres were originally described as vinculin-containing, electron-dense plaques located between the cell membrane and Z-lines in cardiac muscle and in certain types of skeletal muscle (Pardo et al. 1983a; b; Craig and Pardo 1983; Koteliansky and Gneushev 1983; Shear and Bloch 1985). Pardo et al. (1983b) used the term costamere to define these structures because of the rib-like appearance of the plaque when immunostained with anti-vinculin antibody. Based on their location, it was proposed that costameres could provide mechanical integration of the contractile apparatus to the ECM during the contraction-relaxation cycles (Pardo et al. 1983b; Shear and Bloch 1985). This idea was supported by electron microscopic data showing filamentous connection between each Z-line of the peripheral myofibrils and the sarcolemma (Street 1983), and by micrographs showing festooning of the sarcolemma

(Shear and Bloch 1985). We have discovered direct evidence that costameres are also the sites of contraction force transmission to the ECM. When adult rat cardiomyocytes are isolated and plated on laminin, they remodel their cytoskeleton and make focal contact at the costameres (Danowski et al. 1992; Imanaka-Yoshida et al. 1996; 1999). When these cardiomyocytes are cultured on flexible silicone rubber substratum (Harris et al. 1980), characteristic pleat-like wrinkles form and disappear in synchrony with the contraction and relax-ation, respectively (Danowski et al. 1992). Furthermore, microinjection of fluorescent labeled vinculin or α-actinin has enabled us to observe that these pleat-like wrinkles form between adjacent Z-lines, and their presence is always associated with a periodic, "costameric" distribution of vinculin. However, the "costameric" distribution of vinculin is not sufficient for force transmission (Danowski et al. 1992), which suggests that other molecules are necessary to form a complete mechanical linkage. Many molecules present in adhesion plaques, such as talin and integrin $\alpha6\beta1$, are known to show striated distribu-tion patterns similar to costameres (Belkin et al. 1986; Terracio et al. 1991; 1991; Koch-Schneidemann et al. 1994; Belkin et al. 1996; Imanaka-Yoshida et al. 1999). Outside the cell, the ECM protein, laminin, is also localized at costameric attachments sites (Lundgren et al. 1988; Imanaka-Yoshida et al. 1999). These findings reinforce the idea that components of the costamere complex, includ-ing vinculin, talin, integrin $\alpha6\beta1$ and laminin, may function as an attachment complex, capable of transmitting contraction forces generated by the myofib-rils to the extracellular connective tissue, similar to the adhesion plaques of cultured fibroblasts. In fact, anti-$\beta1$ integrin antibody inhibits contraction transmission (Imanaka-Yoshida et al. 1999).

Since the forces transmitted across the sarcolemma of striated muscles are much greater than those of most other nonmuscle cells, the costamere complex may require muscle-specific isoforms. Recently, a cardiac muscle-specific isoform of integrin and its distinctive role have been identified (Belkin et al. 1996; 1997; Baudoin et al. 1998; Brancaccio et al. 1998). Metavinculin, a muscle-specific isoform of vinculin, has been shown to be localized at costameres (Belkin et al. 1988).

Contraction force transmission by cardiomyocytes to the heart interstitial collagen network may be mediated by laminin, because freshly isolated adult cardiomyocytes do not attach to collagen I, but quickly attach to laminin (Lundgren et al. 1985; 1988; Haddad et al. 1988; Koch-Schneidemann et al. 1994) at costameres using integrin $\alpha6\beta1$ (Imanaka-Yoshida et al. 1999). Interestingly, force transmission is not observed at this early stage. Further-more, an anti-$\beta1$ integrin antibody seems to interfere with force transmission only at newly formed costamere attachment sites (unpubl. observ.). These findings suggest that the mature force transmission complex might require modification of the extracellular environment and/or participation of additional intracellular molecules.

7.2.2
Potential Involvement of Dystrophin

Dystroglycan is another laminin receptor(Ibraghimov-Beskrovnaya et al. 1992). As mentioned in the previous section, dystroglycan, together with several dystrophin-associated glycoproteins, including sarcoglycans and syntrophins, forms a complex with dystrophin. Although dystroglycan is an integrin-unrelated adhesion molecule, a possible interaction between integrins and the dystroglycan-dystrophin complex has been suggested (Yoshida et al. 1998). In fact, dystroglycan has been identified in focal adhesions of several cell types (Belkin and Smalheiser 1996). However, in smooth muscle cells, complementary distribution of dystrophin and vinculin has been reported. In striated muscle, one report found dystrophin to be localized at costameres (Straub et al. 1992), but the localization of dystrophin at costamere is controversial (Carpenter et al. 1990; Klietsch et al. 1993; Stevenson et al. 1997). Since it is well known that abnormalities in the laminin-dystroglycan-dystrophin system cause severe muscle dystrophies or cardiomyopathies (reviewed in Ozawa et al. 1998), it is an attractive idea that dystroglycan-dystrophin may function to mediate the force transmission at costammeres. To date, however, there is no concrete evidence of this involvement.

7.2.3
Significance of Costameres in Heart Function

In contrast to skeletal myoblasts, which fuse to each other to form multinucleated myotubes, cardiomyocytes do not fuse. Instead, cardiomyocytes have well-developed cell-cell junctions like epithelial cells, called intercalated. The intercalated disks could be the major sites for contractile force and coordinating forces in the whole heart. In the normal heart, so-called myotendinous junctions are detected only between papillary muscle and valve leaflets. However, the heart consists not only of cardiomyocytes but of abundant connective tissue as well. For the heart to function efficiently as a pump, there should be a mechanical linkage between myofibrils inside the cell and the connective tissue. Costameric attachment may correspond to the myotendinous junctions between cardiomyocytes and the surrounding connective tissue. Thus, abnormalities in these proteins might be expected to cause cardiac failure. The absence of metavinculin has already been reported in a patient with idiopathic dilated cardiomyopathy (Maeda et al. 1997).

8
Conclusion

Contractility generated inside the cells is transmitted to the ECM through cell-matrix adhesion, so that cells not only stabilize their shape and structure, but also reorganize the insoluble ECM. The contractility generated by individual cells is integrated by the ECM, which then transmits it to neighboring cells through the adhesion sites. This mechanotransduction, by activation of the chemical signaling pathway, could directly or indirectly control various cellular events, including regulation of synthesis of ECM, or generation and transmission of cellular contractility itself. The cell contractility and its interaction with the extracellular matrix could provide key regulatory information for tissue organization.

Acknowledgments. The author is grateful to Michael Wise for his careful proof reading of the manuscript. The author also gratefully acknowledges the support by Grants-in-Aid from the Ministry of Education, Science, and Culture of Japan and a Grant-in Aid (1998) from Mie Medical Research Foundation.

References

Arora PD, McCulloch CA (1994) Dependence of collagen remodelling on alpha-smooth muscle actin expression by fibroblasts. J Cell Physiol 159:161–175

Arora PD, Narani N, McCulloch CA (1999) The compliance of collagen gels regulates transforming growth factor-beta induction of alpha-smooth muscle actin in fibroblasts. Am J Pathol 154:871–882

Asaga H, Kikuchi S, Yoshizato K (1991) Collagen gel contraction by fibroblasts requires cellular fibronectin but not plasma fibronectin. Exp Cell Res 193:167–174

Baker LP, Daggett DF, Peng HB (1994) Concentration of pp125 focal adhesion kinase (FAK) at the myotendinous junction. J Cell Sci 107:1485–1497

Baudoin C, Goumans MJ, Mummery C, Sonnenberg A (1998) Knockout and knockin of the beta1 exon D define distinct roles for integrin splice variants in heart function and embryonic development. Genes Dev 12:1202–1216

Belkin AM, Ornatsky OI, Glukhova MA, Koteliansky VE (1988) Immunolocalization of meta-vinculin in human smooth and cardiac muscles. J Cell Biol 107:545–553

Belkin AM, Retta SF, Pletjushkina OY, Balzac F, Silengo L, Fassler R, Koteliansky VE, Burridge K, Tarone G (1997) Muscle beta1D integrin reinforces the cytoskeleton-matrix link: modulation of integrin adhesive function by alternative splicing. J Cell Biol 139:1583–1595

Belkin AM, Smalheiser NR (1996) Localization of cranin (dystroglycan) at sites of cell-matrix and cell-cell contact: recruitment to focal adhesions is dependent upon extracellular ligands. Cell Adhes Commun 4:281–296

Belkin AM, Zhidkova NI, Balzac F, Altruda F, Tomatis D, Maier A, Tarone G, Koteliansky VE, Burridge K (1996) Beta 1D integrin displaces the beta 1A isoform in striated muscles: localization at junctional structures and signaling potential in nonmuscle cells. J Cell Biol 132:211–226

Belkin AM, Zhidkova NI, Koteliansky VE (1986) Localization of talin in skeletal and cardiac muscles. FEBS Lett 200:32–36

Bershadsky A, Chausovsky A, Becker E, Lyubimova A, Geiger B (1996) Involvement of microtubules in the control of adhesion-dependent signal transduction. Curr Biol 6:1279–1289

Berthier C, Blaineau S (1997) Supramolecular organization of the subsarcolemmal cytoskeleton of adult skeletal muscle fibers. A review. Biol Cell 89:413–434

Bozyczko D, Decker C, Muschler J, Horwitz AF (1989) Integrin on developing and adult skeletal muscle. Exp Cell Res 183:72–91

Brancaccio M, Cabodi S, Belkin AM, Collo G, Koteliansky VE, Tomatis D, Altruda F, Silengo L, Tarone G (1998) Differential onset of expression of alpha 7 and beta 1D integrins during mouse heart and skeletal muscle development. Cell Adhes Commun 5:193–205

Burgess ML, Carver WE, Terracio L, Wilson SP, Wilson MA, Borg TK (1994) Integrin-mediated collagen gel contraction by cardiac fibroblasts. Effects of angiotensin II. Circ Res 74:291–298

Burridge K, Chrzanowska-Wodnicka M (1996) Focal adhesions, contractility, and signaling. Ann Rev Cell Dev Biol 12:463–518

Burridge K, Turner CE, Romer LH (1992) Tyrosine phosphorylation of paxillin and pp125FAK accompanies cell adhesion to extracellular matrix: a role in cytoskeletal assembly. J Cell Biol 119:893–903

Byers TJ, Kunkel LM, Watkins SC (1991) The subcellular distribution of dystrophin in mouse skeletal, cardiac, and smooth muscle. J Cell Biol 115:411–421

Campbell KP (1995) Three muscular dystrophies: loss of cytoskeleton-extracellular matrix linkage. Cell 80:675–679

Carpenter S, Karpati G, Zubrzycka-Gaarn E, Bulman DE, Ray PN, Worton RG (1990) Dystrophin is localized to the plasma membrane of human skeletal muscle fibers by electron-microscopic cytochemical study. Muscle Nerve 13:376–380

Carver W, Nagpal ML, Nachtigal M, Borg TK, Terracio L (1991) Collagen expression in mechanically stimulated cardiac fibroblasts. Circ Res 69:116–122

Carver W, Molano I, Reaves TA, Borg TK, Terracio L (1995) Role of the alpha 1 beta 1 integrin complex in collagen gel contraction in vitro by fibroblasts. J Cell Physiol 165:425–437

Chicurel ME, Chen CS, Ingber DE (1997) Cellular control lies in the balance of forces. Curr Opin Cell Biol 10:232–239

Chiquet-Ehrismann R, Tannheimer M, Koch M, Brunner A, Spring J, Martin D, Baumgartner S, Chiquet M (1994) Tenascin-C expression by fibroblasts is elevated in stressed collagen gels. J Cell Biol 127:2093–2101

Choquet D, Felsenfeld DP, Sheetz MP (1997) Extracellular matrix rigidity causes strengthening of integrin-cytoskeleton linkages. Cell 88:39–48

Chrzanowska-Wodnicka M, Burridge K (1994) Tyrosine phosphorylation is involved in reorganization of the actin cytoskeleton in response to serum or LPA stimulation. J Cell Sci 107:3643–3654

Chrzanowska-Wodnicka M, Burridge K (1996) Rho-stimulated contractility drives the formation of stress fibers and focal adhesions. J Cell Biol 133:1403–1415

Craig SW, Pardo JV (1983) Gamma actin, spectrin, and intermediate filament proteins colocalize with vinculin at costameres, myofibril-to-sarcolemma attachment sites. Cell Motil 3:449–462

Crowley E, Horwitz AF (1995) Tyrosine phosphorylation and cytoskeletal tension regulate the release of fibroblast adhesions. J Cell Biol 131:525–537

Danowski BA (1989) Fibroblast contractility and actin organization are stimulated by microtubule inhibitors. J Cell Sci 93:255–266

Danowski BA (1998) Microtubule dynamics in serum-starved and serum-stimulated Swiss 3T3 mouse fibroblasts: implications for the relationship between serum-induced contractility and microtubules. Cell Motil Cytoskeleton 40:1–12

Danowski BA, Harris AK (1988) Changes in fibroblast contractility, morphology, and adhesion in response to a phorbol ester tumor promoter. Exp Cell Res 177:47–59

Danowski BA, Imanaka-Yoshida K, Sanger JM, Sanger JW (1992) Costameres are sites of force transmission to the substratum in adult rat cardiomyocytes. J Cell Biol 118:1411–1420

Darby I, Skalli O, Gabbiani G (1990) Alpha-smooth muscle actin is transiently expressed by myofibroblasts during experimental wound healing. Lab Invest 63:21–29

Eckes B, Mauch C, Hüppe G, Krieg T (1993) Downregulation of collagen synthesis in fibroblasts within three-dimensional collagen lattices involves transcriptional and posttranscriptional mechanisms. FEBS Lett 318:129–133

Engvall E, Earwicker D, Haaparanta T, Ruoslahti E, Sanes JR (1990) Distribution and isolation of four laminin variants; tissue-restricted distribution of heterotrimers assembled from five different subunits. Cell Regul 1:731–740

Enomoto T (1996) Microtubule disruption induces the formation of actin stress fibers and focal adhesions in cultured cells: possible involvement of the rho signal cascade. Cell Struct Funct 21:317–326

Ezzell RM, Goldmann WH, Wang N, Parasharama N, Ingber DE (1997) Vinculin promotes cell spreading by mechanically coupling integrins to the cytoskeleton. Exp Cell Res 231:14–26

Flinn HM, Ridley AJ (1996) Rho stimulates tyrosine phosphorylation of focal adhesion kinase, p130 and paxillin. J Cell Sci 109:1133–1141

Franke RP, Gräfe M, Schnittler H, Seiffge D, Mittermayer C, Drenckhahn D (1984) Induction of human vascular endothelial stress fibres by fluid shear stress. Nature 307:648–649

Frenette J, Tidball JG (1998) Mechanical loading regulates expression of talin and its mRNA, which are concentrated at myotendinous junctions. Am J Physiol 275:C818–825

Glass WF2d, Kreisberg JI (1993) Regulation of integrin-mediated adhesion at focal contacts by cyclic AMP. J Cell Physiol 157:296–306

Goldmann WH, Galneder R, Ludwig M, Xu W, Adamson ED, Wang N, Ezzell RM (1998) Differences in elasticity of vinculin-deficient F9 cells measured by magnetometry and atomic force microscopy. Exp Cell Res 239:235–242

Grinnell F (1994) Fibroblasts, myofibroblasts, and wound contraction. J Cell Biol 124:401–404

Grinnell F, Ho CH, Lin YC, Skuta G (1999) Differences in the regulation of fibroblast contraction of floating versus stressed collagen matrices. J Biol Chem 274:918–923

Gullberg D, Tingström A, Thuresson AC, Olsson L, Terracio L, Borg TK, Rubin K (1990) Beta 1 integrin-mediated collagen gel contraction is stimulated by PDGF. Exp Cell Res 186:264–272

Haddad J, Decker ML, Hsieh LC, Lesch M, Samarel AM, Decker RS (1988) Attachment and maintenance of adult rabbit cardiac myocytes in primary cell culture. Am J Physiol 255:C19–27

Halliday NL, Tomasek JJ (1995) Mechanical properties of the extracellular matrix influence fibronectin fibril assembly in vitro. Exp Cell Res 217:109–117

Hamasaki K, Mimura T, Furuya H, Morino N, Yamazaki T, Komuro I, Yazaki Y, Nojima Y (1995) Stretching mesangial cells stimulates tyrosine phosphorylation of focal adhesion kinase pp125FAK. Biochem Biophys Res Commun 212:544–549

Harris AK (1994) Locomotion of tissue culture cells considered in relation to ameboid locomotion. Int Rev Cytol 150:35–68

Harris AK, Wild P, Stopak D (1980) Silicone rubber substrata: a new wrinkle in the study of cell locomotion. Science 208:177–179

Harris AK, Stopak D, Wild P (1981) Fibroblast traction as a mechanism for collagen morphogenesis. Nature 290:249–251

Hatamochi A, Aumailley M, Mauch C, Chu ML, Timpl R, Krieg T (1989) Regulation of collagen VI expression in fibroblasts. Effects of cell density, cell-matrix interactions, and chemical transformation. J Biol Chem 264:3494–3499

Hoffman EP, Brown RHJ, Kunkel LM (1987) Dystrophin: the protein product of the Duchenne muscular dystrophy locus. Cell 51:919–928

Ibraghimov-Beskrovnaya O, Ervasti JM, Leveille CJ, Slaughter CA, Sernett SW, Campbell KP (1992) Primary structure of dystrophin-associated glycoproteins linking dystrophin to the extracellular matrix. Nature 355:696–702

Imanaka-Yoshida K, Danowski BA, Sanger JM, Sanger JW (1996) Living adult rat cardiomyocytes in culture: evidence for dissociation of costameric distribution of vinculin from costameric distributions of attachments. Cell Motil Cytoskeleton 33:263–275

Imanaka-Yoshida K, Enomoto-Iwamoto M, Yoshida T, Sakakura T (1999) Vinculin, talin, integrin $\alpha6\beta1$ and laminin can serve as components of attachment complex mediating contraction force transmission from cardiomyocytes to extracellular matrix. Cell Motil Cytoskeleton 42:1–11

Ingber DE (1993) Cellular tensegrity: defining new rules of biological design that govern the cytoskeleton. J Cell Sci 104:613–627

Ingber DE (1997) Tensegrity: the architectural basis of cellular mechanotransduction. Annu Rev Physiol 59:575–599

Ingber DE, Dike L, Hansen L, Karp S, Liley H, Maniotis A, McNamee H, Mooney D, Plopper G, Sims J (1994) Cellular tensegrity: exploring how mechanical changes in the cytoskeleton regulate cell growth, migration, and tissue pattern during morphogenesis. Int Rev Cytol 150:173–224

Jockusch BM, Bubeck P, Giehl K, Kroemker M, Moschner J, Rothkegel M, Rüdiger M, Schlüter K, Stanke G, Winkler J (1995) The molecular architecture of focal adhesions. Ann Rev Cell Dev Biol 11:379–416

Johnson RP, Craig SW (1994) An intramolecular association between the head and tail domains of vinculin modulates talin binding. J Biol Chem 269:12611–12619

Kimura K, Ito M, Amano M, Chihara K, Fukata Y, Nakafuku M, Yamamori B, Feng J, Nakano T, Okawa K, Iwamatsu A, Kaibuchi K (1996) Regulation of myosin phosphatase by Rho and Rho-associated kinase (Rho-kinase). Science 273:245–248

Klein CE, Dressel D, Steinmayer T, Mauch C, Eckes B, Krieg T, Bankert RB, Weber L (1991) Integrin alpha 2 beta 1 is upregulated in fibroblasts and highly aggressive melanoma cells in three-dimensional collagen lattices and mediates the reorganization of collagen I fibrils. J Cell Biol 115:1427–1436

Klietsch R, Ervasti JM, Arnold W, Campbell KP, Jorgensen AO (1993) Dystrophin-glycoprotein complex and laminin colocalize to the sarcolemma and transverse tubules of cardiac muscle. Circ Res 72:349–360

Koch-Schneidemann S, Gehr P, Rutishauser B, Eppenberger HM (1994) Attachment of adult rat cardiomyocytes (ARC) on laminin and two laminin fragments. J Struct Biol 113:107–116

Kolega J (1986) Effects of mechanical tension on protrusive activity and microfilament and intermediate filament organization in an epidermal epithelium moving in culture. J Cell Biol 102:1400–1411

Kolodney MS, Elson EL (1995) Contraction due to microtubule disruption is associated with increased phosphorylation of myosin regulatory light chain. Proc Natl Acad Sci USA 92:10252–10256

Koteliansky VE, Gneushev GN (1983) Vinculin localization in cardiac muscle. FEBS Lett 159:158–160

Lamb NJ, Fernandez A, Conti MA, Adelstein R, Glass DB, Welch WJ, Feramisco JR (1988) Regulation of actin microfilament integrity in living nonmuscle cells by the cAMP-dependent protein kinase and the myosin light chain kinase. J Cell Biol 106:1955–1971

Lambert CA, Soudant EP, Nusgens BV, Lapière CM (1992) Pretranslational regulation of extracellular matrix macromolecules and collagenase expression in fibroblasts by mechanical forces. Lab Invest 66:444–451

Lampugnani MG, Giorgi M, Gaboli M, Dejana E, Marchisio PC (1990) Endothelial cell motility, integrin receptor clustering, and microfilament organization are inhibited by agents that increase intracellular cAMP. Lab Invest 63:521–531

Lauffenburger DA, Horwitz AF (1996) Cell migration: a physically integrated molecular process. Cell 84:359–369

Lee J, Leonard M, Oliver T, Ishihara A, Jacobson K (1994) Traction forces generated by locomoting keratocytes. J Cell Biol 127:1957–1964

Liu BP, Chrzanowska-Wodnicka M, Burridge K (1998) Microtubule depolymerization induces stress fibers, focal adhesions, and DNA synthesis via the GTP-binding protein Rho. Cell Adhes Commun 5:249–255

Lundgren E, Terracio L, Mårdh S, Borg TK (1985) Extracellular matrix component influence the survival of adult cardiac myocytes in vitro. Exp Cell Res 158:371–381

Lundgren E, Gullberg D, Rubin K, Borg TK, Terracio MJ, Terracio L (1988) In vitro studies on cardiac myocytes: Attachment and biosynthesis of collagen type IV and laminin. J Cell Physiol 136:43–53

Maeda M, Holder E, Lowes B, Valent S, Bies RD (1997) Dilated cardiomyopathy associated with deficiency of the cytoskeletal protein metavinculin. Circulation 95:17–20

Maniotis AJ, Chen CS, Ingber DE (1997) Demonstration of mechanical connections between integrins, cytoskeletal filaments, and nucleoplasm that stabilize nuclear structure. Proc Natl Acad Sci USA 94:849–854

Matsumura K, Campbell KP (1994) Dystrophin-glycoprotein complex: its role in the molecular pathogenesis of muscular dystrophies. Muscle Nerve 17:2–15

Mitchison TJ, Cramer LP (1996) Actin-based cell motility and cell locomotion. Cell 84:371–379

Mochitate K, Pawelek P, Grinnell F (1991) Stress relaxation of contracted collagen gels: disruption of actin filament bundles, release of cell surface fibronectin, and down-regulation of DNA and protein synthesis. Exp Cell Res 193:198–207

Murphy-Ullrich JE, Höök M (1989) Thrombospondin modulates focal adhesions in endothelial cells. J Cell Biol 109:1309–1319

Murphy-Ullrich JE, Lightner VA, Aukhil I, Yan YZ, Erickson HP, Höök M (1991) Focal adhesion integrity is downregulated by the alternatively spliced domain of human tenascin [published erratum appears in J Cell Biol 1992 Feb;116(3):833]. J Cell Biol 115:1127–1136

Murphy-Ullrich JE, Pallero MA, Boerth N, Greenwood JA, Lincoln TM, Cornwell TL (1996) Cyclic GMP-dependent protein kinase is required for thrombospondin and tenascin mediated focal adhesion disassembly. J Cell Sci 109:2499–2508

Ohashi T, Kiehart DP, Erickson HP (1999) Dynamics and elasticity of the fibronectin matrix in living cell culture visualized by fibronectin-green fluorescent protein. Proc Natl Acad Sci USA 96:2153–2158

Oliver T, Dembo M, Jacobson K (1995) Traction forces in locomoting cells. Cell Motil Cytoskeleton 31:225–240

Ozawa E, Noguchi S, Mizuno Y, Hagiwara Y, Yoshida M (1998) From dystrophinopathy to sarcoglycanopathy: evolution of a concept of muscular dystrophy. Muscle Nerve 21:421–438

Pardo JV, Siliciano JD, Craig SW (1983a) Vinculin is a component of an extensive network of myofibril-sarcolemma attachment regions in cardiac muscle fibers. J Cell Biol 97:1081–1088

Pardo JV, Siliciano JD, Craig SW (1983b) A vinculin-containing cortical lattice in skeletal muscle: transverse lattice elements (costameres) mark sites of attachment between myofibrils and sarcolemma. Proc Natl Acad Sci USA 80:1008–1012

Petroll WM, Cavanagh HD, Barry P, Andrews P, Jester JV (1993) Quantitative analysis of stress fiber orientation during corneal wound contraction. J Cell Sci 104:353–363

Racine-Samson L, Rockey DC, Bissell DM (1997) The role of alpha1beta1 integrin in wound contraction. A quantitative analysis of liver myofibroblasts in vivo and in primary culture. J Biol Chem 272:30911–30917

Riikonen T, Koivisto L, Vihinen P, Heino J (1995) Transforming growth factor-beta regulates collagen gel contraction by increasing alpha 2 beta 1 integrin expression in osteogenic cells. J Biol Chem 270:376–382

Schiro JA, Chan BM, Roswit WT, Kassner PD, Pentland AP, Hemler ME, Eisen AZ, Kupper TS (1991) Integrin alpha 2 beta 1 (VLA-2) mediates reorganization and contraction of collagen matrices by human cells. Cell 67:403–410

Schliwa M, Nakamura T, Porter KR, Euteneuer U (1984) A tumor promoter induces rapid and coordinated reorganization of actin and vinculin in cultured cells. J Cell Biol 99:1045–1059

Schoenwaelder SM, Burridge K (1999) Bidirectional signaling between the cytoskeleton and integrins. Current Opinion In Cell Biology 11:274–286

Seufferlein T, Rozengurt E (1994) Sphingosine induces p125FAK and paxillin tyrosine phosphorylation, actin stress fiber formation, and focal contact assembly in Swiss 3T3 cells. J Biol Chem 269:27610–27617

Shear CR, Bloch RJ (1985) Vinculin in subsarcolemmal densities in chicken skeletal muscle: localization and relationship to intracellular and extracellular structures. J Cell Biol 101:240–256

Sheetz MP (1994) Cell migration by graded attachment to substrates and contraction. Semin Cell Biol 5:149–155

Shyy JY, Chien S (1997) Role of integrins in cellular responses to mechanical stress and adhesion. Curr Opin Cell Biol 9:707–713

Singer II, Kawka DW, Kazazis DM, Clark RA (1984) In vivo co-distribution of fibronectin and actin fibers in granulation tissue: immunofluorescence and electron microscope studies of the fibronexus at the myofibroblast surface. J Cell Biol 98:2091–2106

Small JV, Fürst DO, Thornell LE (1992) The cytoskeletal lattice of muscle cells. Eur J Biochem 208:559–572

Stevenson S, Rothery S, Cullen MJ, Severs NJ (1997) Dystrophin is not a specific component of the cardiac costamere. Circ Res 80:269–280

Stopak D, Harris AK (1982) Connective tissue morphogenesis by fibroblast traction. I. Tissue culture observations. Dev Biol 90:383–398

Stopak D, Wessells NK, Harris AK (1985) Morphogenetic rearrangement of injected collagen in developing chicken limb buds. Proc Natl Acad Sci USA 82:2804–2808

Straub V, Bittner RE, Léger JJ, Voit T (1992) Direct visualization of the dystrophin network on skeletal muscle fiber membrane. J Cell Biol 119:1183–1191

Street SF (1983) Lateral transmission of tension in frog myofibers: a myofibrillar network and transverse cytoskeletal connections are possible transmitters. J Cell Physiol 114:346–364

Swasdison S, Mayne R (1989) Location of the integrin complex and extracellular matrix molecules at the chicken myotendinous junction. Cell Tissue Res 257:537–543

Terracio L, Simpson DG, Hilenski L, Carver W, Decker RS, Vinson N, Borg TK (1990) Distribution of vinculin in the Z-disk of striated muscle: analysis by laser scanning confocal microscopy. J Cell Physiol 145:78–87

Terracio L, Rubin K, Gullberg D, Balog E, Carver W, Jyring R, Borg TK (1991) Expression of collagen binding integrins during cardiac development and hypertrophy. Circ Res 68:734–744

Tidball JG, Daniel TL (1986) Myotendinous junctions of tonic muscle cells: structure and loading. Cell Tissue Res 245:315–322

Tinsley JM, Blake DJ, Zuellig RA, Davies KE (1994) Increasing complexity of the dystrophin-associated protein complex. Proc Natl Acad Sci USA 91:8307–8313

Tomasek JJ, Akiyama SK (1992) Fibroblast-mediated collagen gel contraction does not require fibronectin-alpha 5 beta 1 integrin interaction. Anat Rec 234:153–160

Tomasek JJ, Haaksma CJ (1991) Fibronectin filaments and actin microfilaments are organized into a fibronexus in Dupuytren's diseased tissue. Anat Rec 230:175–182

Tomasek JJ, Haaksma CJ, Eddy RJ, Vaughan MB (1992) Fibroblast contraction occurs on release of tension in attached collagen lattices: dependency on an organized actin cytoskeleton and serum. Anat Rec 232:359–368

Trächslin J, Koch M, Chiquet M (1999) Rapid and reversible regulation of collagen XII expression by changes in tensile stress. Exp Cell Res 247:320–328

Wang N, Butler JP, Ingber DE (1993) Mechanotransduction across the cell surface and through the cytoskeleton. Science 260:1124–1127

Weekes J, Barry ST, Critchley DR (1996) Acidic phospholipids inhibit the intramolecular association between the N- and C-terminal regions of vinculin, exposing actin-binding and protein kinase C phosphorylation sites. Biochem J 314:827–832

Yoshida T, Pan Y, Hanada H, Iwata Y, Shigekawa M (1998) Bidirectional signaling between sarcoglycans and the integrin adhesion system in cultured L6 myocytes. Journal Of Biol Chem 273:1583–1590

Zhang Q, Checovich WJ, Peters DM, Albrecht RM, Mosher DF (1994) Modulation of cell surface fibronectin assembly sites by lysophosphatidic acid. J Cell Biol 127:1447–1459

Zhang Q, Magnusson MK, Mosher DF (1997) Lysophosphatidic acid and microtubule-destabilizing agents stimulate fibronectin matrix assembly through Rho-dependent actin stress fiber formation and cell contraction. Mol Biol Cell 8:1415–1425

Zhong C, Chrzanowska-Wodnicka M, Brown J, Shaub A, Belkin AM, Burridge K (1998) Rho-mediated contractility exposes a cryptic site in fibronectin and induces fibronectin matrix assembly. J Cell Biol 141:539–551

Role of Focal Adhesion Kinase in Signaling by the Extracellular Matrix

Ji-He Zhao and Jun-Lin Guan[1]

1
Introduction

Cellular interactions with extracellular matrix (ECM) play important roles in many biological processes such as embryonic development and morphogenesis, wound healing, and malignant transformation. The major cell surface receptors for ECM are the integrin family cell adhesion molecules which are composed of noncovalently associated α and β subunits (Hynes 1992). Both α and β subunits are transmembrane proteins containing short cytoplasmic domains that can interact directly or indirectly with cytoskeletal proteins such as talin, vinculin, α-actinin, and filamin (Burridge et al. 1992; Hynes 1992). Besides serving as a linker between ECM and cytoskeleton, integrins have been shown to mediate biochemical signal transduction across the plasma membrane to regulate various cellular functions (Juliano and Haskill 1993; Clark and Brugge 1995; Schwartz et al. 1995). Integrin-mediated cell adhesion can regulate gene expression, intracellular pH and calcium, phospholipid metabolites, small GTPase, protein serine/threonine kinases, and protein tyrosine phosphorylation. Since integrins have short cytoplasmic domains with no enzymatic activities, it is believed that coupling of the integrin cytoplasmic domains with cytoplasmic tyrosine kinases, phosphatases, and adaptor molecules are critical in initiating multiple intracellular signaling pathways mediated by integrins (Clark and Brugge 1995; Juliano and Haskill 1993; Schwartz et al. 1995).

Focal adhesion kinase (FAK) is a cytoplasmic protein tyrosine kinase which has been implicated in playing an important role in integrin-mediated signal transduction pathways (Clark and Brugge 1995; Schwartz et al. 1995; Parsons 1996). FAK becomes activated and tyrosine phosphorylated in integrin-mediated cell adhesion and is colocalized with integrins and other cytoskeletal proteins in focal contacts in a variety of adherent cells (Schwartz et al. 1995). cDNAs encoding FAK from avian (Schaller et al. 1992), murine (Hanks et al. 1992), human (Andre and Becker-Andre 1993; Whitney et al. 1993), and *Xenopus* (Hens and DeSimone 1995) have been described. The comparison of

[1] Cancer Biology Laboratories, Department of Molecular Medicine, College of Veterinary Medicine, Cornell University, Ithaca, New York 14853, USA

Progress in Molecular and Subcellular Biology, Vol. 25
A. Macieira-Coelho (Ed.)

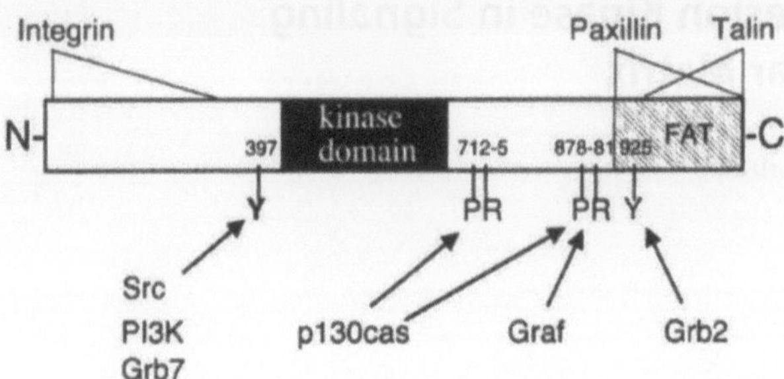

Fig. 1. FAK structure and associated proteins. The kinase domain is centrally located. Putative integrin-binding region within the N-terminal domain and binding regions for paxillin and talin within the *FAT* (focal adhesion targeting) sequence in the C-terminal domain are shown above the FAK diagram. Tyrosine phosphorylation sites (*Y*) and proline-rich (*PR*) regions are indicated below the FAK diagram. The SH2- and SH3-containing proteins, which bind to the correspondent *Y* and *PR* regions, respectively, are also shown below the FAK diagram

deduced amino acid sequences from the cDNAs reveals a remarkably high degree of identity (~90%), suggesting that the cellular function of FAK is highly conserved in evolution. FAK (Hanks et al. 1992; Schaller et al. 1992) and a closely related tyrosine kinase, Pyk2, (also termed CAKβ, RAFTK, FAK2; Avraham et al. 1995; Lev et al. 1995; Sasaki et al. 1995; Herzog et al. 1996) have a unique structural organization among the cytoplasmic protein tyrosine kinases. The central catalytic domain of the kinases are flanked by large N-terminal and C-terminal domains which do not contain Src-homology 2 (SH2) or Src-homology 3 (SH3) domain. Instead, the flanking noncatalytic domains have phosphotyrosine- and proline-rich sequences which serve as potential binding sites for other SH2 and SH3 domain-containing proteins, respectively. Within the C-terminal domain of FAK, a focal adhesion targeting (FAT) sequence has been identified, which is responsible for FAK localization to focal adhesions (Hildebrand et al. 1993). These key structural features of FAK and its binding sites for other cellular proteins (see below) are illustrated in Fig. 1.

Discovery of FAK and its regulation by integrins in the early 1990s triggered great interest in studies of signal transduction by ECM (Guan et al. 1991; Kornberg et al. 1991; Hanks et al. 1992; Schaller et al. 1992). Research in the past several years has clearly established a critical role for FAK in integrin-mediated signaling by ECM. This chapter will summarize our current under-standing of FAK and its associated proteins in signal transduction pathways initiated by cellular interaction with ECM. We will first discuss current studies on the mechanisms of FAK activation by integrins. We will then focus on the interactions of the activated FAK with other cellular signaling molecules, which trigger several downstream signaling pathways. Finally, we will discuss recent studies to examine the role of these signaling pathways in the context of cellular functions regulated by integrins.

2
FAK Activation by Integrin

The critical role of integrins in FAK activation during cell adhesion to ECM was first suggested from initial studies on FAK in the early 1990s. Tyrosine phosphorylation of FAK was strongly induced when fibroblasts were replated on integrin ligand fibronectin (FN) or on antiintegrin antibodies (Guan et al. 1991; Guan and Shalloway 1992) or when the cell surface integrins of KB carcinoma cells were clustered with integrin antibodies (Kornberg et al. 1991, 1992). Furthermore, FAK has been shown to colocalize with integrins in focal adhesions in a variety of adherent cells. Although the mechanisms remain incompletely understood, two models (see below) have been suggested which involve FAK interactions either directly or indirectly with the cytoplasmic domains of integrins.

2.1
FAK Interaction with Integrin

Integrin β1 cytoplasmic domain has been shown to play a critical role in both focal adhesion formation (Solowska et al. 1989) and induction of FAK phosphorylation (Guan et al. 1991; Akiyama et al. 1994; Lukashev et al. 1994). These observations have led to the proposal that FAK is activated by directly interacting with the cytoplasmic domain of integrin β1. This model is further supported by the findings that FAK could bind directly to a peptide that mimics the membrane proximal region of cytoplasmic tail of integrin β1 in vitro (Schaller et al. 1995). However, there has been no further evidence to support a direct interaction of FAK with integrins by in vivo or other in vitro binding assays. In addition, the membrane distal, instead of proximal, region of integrin β1 cytoplasmic tail has been demonstrated to be crucial for inducing FAK phosphorylation in vivo (Guan et al. 1991). Therefore, (an)other mediator protein(s) is likely required for FAK activation by integrins.

2.2
FAK Interaction with Cytoskeletal Proteins Talin, Paxillin and Tensin

The alternative model suggests that FAK is activated by integrins through its indirect association with integrin cytoplasmic domains via binding to cytoskeletal proteins colocalized in focal adhesions. The integrity of actin cytoskeleton is critical for focal adhesion formation (Horwitz et al. 1986) as well as for FAK phosphorylation and activation since selective disruption of cytoskeleton with cytochalasin D specifically blocked phosphorylation and activation of FAK (Burridge et al. 1992; Guan and Shalloway 1992; Sinnett-Smith et al. 1993). Several lines of evidence suggested the cytoskeletal protein talin as a mediator of FAK activation by integrins. First, talin has been shown to bind directly to integrins as well as FAK in vivo (Horwitz et al. 1986; Chen et al. 1995). Second, talin binding to FAK did not depend upon FAK phosphorylation and activation, suggesting that this binding occurred prior to FAK activation and phosphorylation by integrins (Guan 1997). Third, the membrane distal region of β1 integrin was required for both integrin activation of FAK (Guan et al. 1991) and talin binding to β1 integrin as shown by protein interaction assays (Chen et al. 1995) or antiintegrin antibody-coated microbeads-induced integrin aggregation assays (Lewis and Schwartz 1995). Finally, recent study demonstrated that differential phopshorylation of Pyk2 and FAK in cell adhesion correlated to their binding to talin (Zheng et al. 1998).

Another cytoskeletal protein paxillin is also colocalized to focal adhesions with FAK and integrins and has been shown to directly associate with FAK in vivo (Turner and Miller 1994; Schaller and Parsons 1995). Since the paxillin binding sequence on FAK overlaps well with the FAT sequence (Fig. 1; Hildebrand et al. 1995), it has been proposed that paxillin may play a key role in FAK localization in focal adhesions and its activation by integrins. However, one FAK deletion mutant has been found to localize in focal adhesions although it does not bind to paxillin (Tachibana et al. 1995). Conversely, Pyk2 has been shown to associate with paxillin while not being localized to focal adhesions in several cell lines (Schaller and Sasaki 1997; Xiong et al. 1998; Zheng et al. 1998). By using an antiintegrin antibody-coated microbeads-induced integrin aggregation assays, Miyamoto et al. (1995) have shown that the cytoskeletal protein tensin could be coaggregated with phosphorylated FAK and integrins independent of the presence of other known cytoskeletal proteins such as talin and paxillin. Together, these results suggest a complex model of FAK activation by integrins, which may involve cooperative interactions of FAK with integrins as well as a number of cytoskeletal proteins.

3
FAK Downstream Pathways

Like other intracellular signaling molecules, FAK has been shown to transduce biochemical signals through direct interactions with a number of downstream signaling molecules. Although it does not contain SH2 and SH3 domains as other cytoplasmic tyrosine kinases, FAK contains phosphotyrosine motifs that serve as binding sites for SH2-containing signaling proteins and proline-rich regions that are binding sites for SH3-containing proteins. FAK Y397 (within YAEI motif) is the major autophosphorylation site in vitro as well as upon integrin-mediated cell adhesion in vivo (Chan et al. 1994; Schaller et al. 1994). The phosphorylated Y397 has been shown to be a binding site for SH2 domain of Src (Schaller et al. 1994; Eide et al. 1995; Xing et al. 1994), PI 3-kinase (phosphatidylinositol 3-kinase) (Chen et al. 1996a), and Grb7 adapter protein (Han and Guan 1999). Y925 of FAK has been shown to be a phosphorylation site by Src and a binding site for the SH2 domain of Grb2 (Schlaepfer et al. 1994). The proline-rich regions in the C-terminal domain of FAK have been shown to associate with SH3-domain of p130Cas (Crk-associated substrate) (Ishino et al. 1995; Polte and Hanks 1995; Harte et al. 1996) and Graf (GTPase regulator associated with FAK) (Hildebrand et al. 1996). These interactions will be discussed in some detail in proportion to our understanding of their potential role in FAK-mediated signal transduction pathways initiated by ECM.

3.1
FAK Interaction with Src

Src family tyrosine kinases were the first identified FAK-associated proteins. The stable association of FAK with Src family members (e.g., v-Src, c-Src and c-Fyn) has been demonstrated using a number of in vivo and in vitro binding assays (Schaller et al. 1994; Schlaepfer et al. 1994; Cobb et al. 1994; Eide et al. 1995; Xing et al. 1994; Cary et al. 1996; Zhao et al. 1998). The association of FAK with c-Src is detected in cells replated on FN but not poly-lysine, indicating that the association of FAK with c-Src is downstream of FAK activation and autophosphorylation and that the integrin signaling alone is sufficient to mediate association of FAK with c-Src without involvement of any growth factors (Schlaepfer et al. 1994). The intimate relationship between FAK and Src family tyrosine kinases is not surprising in that FAK was originally identified as a major protein with increased tyrosine phosphorylation in v-Src-transformed cells (Kanner et al. 1990; Schaller et al. 1992). Interestingly, increased FAK phosphorylation in v-Src-transformed cells did not depend on cell attachment on ECM, suggesting that the integrin-mediated signaling through FAK in the transformed cells was constitutively activated, which may be required for the anchorage-independent growth of transformed cells (Guan and Shalloway 1992).

Src associates with FAK through direct binding of its SH2 domain to the major autophosphorylation site Y397 on FAK (Schaller et al. 1994; Schaller and Parsons 1994; Eide et al. 1995; Xing et al. 1994). Recent studies also showed that a simultaneous binding of Src SH3 domain to a proline-rich region immediately upstream of this Y397 might contribute to maximal association of Src with FAK (Thomas et al. 1998). The FAK/Src association is believed to play an important role in recruiting Src to focal adhesions and in reciprocal activation of these two kinases (Guan 1997). The association may lead to activation of both kinases by protecting them from inhibitory protein tyrosine kinase (Csk for Src) (Cooper and Howell 1993) or phosphotases (e.g., PTP1B, PTP-PEST or PTEN for FAK) (Liu et al. 1998; Tamura et al. 1998; Angers-Laustau et al. 1999).

Src binding to FAK has been proposed to allow Src to phosphorylate several other tyrosine residues on FAK, which leads to further activation of FAK and allows FAK binding to additional SH2 domain-containing signaling proteins. Src phosphorylation of Y576 and Y577 in the kinase domain of FAK has been shown to increase FAK catalytic activity (Calalb et al. 1995). The potential functions of Src phosphorylation of Y407 or Y861 are not clear, but are suggested to mediate binding to SH2 domain-containing molecules (Calalb et al. 1995, 1996). Finally, phosphorylation of Y925 of FAK by Src has been shown to create a binding site for the SH2 domain of Grb2 (Schlaepfer et al. 1994; Schlaepfer and Hunter 1996). Grb2 is an adapter molecule that connects activated receptor tyrosine kinases to Ras/Erk MAP kinase signaling pathway by binding to Sos, a guanine nucleotide exchange factor for Ras (Buday and Downward 1993; Chardin et al. 1993; Egan et al. 1993). Therefore, FAK association with Grb2/Sos in response to cell adhesion might provide a mechanism by which the Erk signaling pathway is activated in cell adhesion to ECM.

The Src/FAK complex has also been suggested to phosphorylate several other proteins including cytoskeletal protein paxillin (Burridge et al. 1992; Turner et al. 1993), tensin (Bockholt and Burridge 1993), and p130cas (Nojima et al. 1995; Vuori and Ruoslahti 1995). Both paxillin and p130cas can be phosphorylated by FAK or Src in vitro (Turner et al. 1993; Sabe et al. 1994; Sakai et al. 1994). However, their phosphorylation in vivo requires the FAK autophosphorylation and Src binding site Y397 (Schaller and Parsons 1995; Vuori et al. 1996). Interestingly, p130cas phosphorylation is diminished in Src−/− cells, but increased in Csk−/− cells in which Src activity is elevated (Vuori et al. 1996). Therefore, it is likely that the FAK/Src complex is responsible for phosphorylating paxillin and p130cas in vivo. Besides potential effects on regulation of cytoskeleton, tyrosine phosphorylation of p130cas and paxillin by FAK/Src signaling complexes generates binding sites for other SH2 domain-containing proteins. Crk, an adapter protein with one SH2 and two SH3 domains, has been shown to bind paxillin in vitro (Birge et al. 1993). Recent mapping studies suggest that the major phosphorylation sites of paxillin, Y31 and Y118, conform to the Crk SH2 binding consensus (Schaller and Parsons 1995). The SH3 domains of Crk could bind to Sos and to another guanine nucleotide exchange factor for Ras, C3G (Matsuda et al. 1994; Tanaka et al. 1994). Likewise, p130cas has also been shown to bind Crk in a cell adhesion-dependent manner (Vuori et al. 1996). Therefore, tyrosine phosphorylation of p130cas and paxillin by FAK/Src complex could provide redundant mechanism by which Ras/MAP kinase pathway is activated in signal transduction by integrins (Schlaepfer et al. 1997).

3.2
FAK Interaction with PI 3-Kinase

In addition to Src, the major autophosphorylation site Y397 on FAK also serves as a binding site for PI 3-kinase (Chen et al. 1996a). PI 3-kinase is composed of a 110-kDa catalytic subunit (p110) and an 85-kDa regulatory subunit (p85) containing SH2 and SH3 domains. Association of FAK with PI 3-kinase in response to integrin activation has been demonstrated in both platelets (Guinebault et al. 1995) and fibroblasts (Chen and Guan 1994). In vitro binding assays demonstrated that the association of FAK with PI 3-kinase is through direct binding of FAK Y397 and the SH2 domain of the p85 subunit of PI 3-

kinase. This is consistent with the fact that FAK autophosphorylation upon cell adhesion stimulates FAK/PI 3-kinase association (Chen and Guan 1994). In vivo coimmunoprecipitation experiments confirmed the requirement of autophosphorylation site Y397 for FAK binding to PI 3-kinase (Chen et al. 1996a). It has also been shown that cell adhesion increased tyrosine phosphorylation of p85 in vivo and that p85 can be phosphorylated directly by FAK in vitro (Chen and Guan 1994). Together these results suggest that PI 3-kinase may be another downstream effector of FAK.

The critical role of Y397 of FAK in regulation of cell proliferation, apoptosis, and migration (as described below) suggests that PI 3-kinase binding to FAK may be involved in any or all of these cellular functions. Both PI 3-kinase and its downstream target, PKB/Akt, have been shown to be stimulated upon adhesion of epithelial cells to ECM (Khwaja et al. 1997; King et al. 1997). Interestingly, expression of the constitutively activated PI 3-kinase or PKB/Akt, as well as a membrane-anchored, constitutively active form of FAK, has been shown to protect MDCK epithelial cells from anoikis, apoptosis induced by the cell detachment from ECM (Frisch et al. 1996a; Kennedy et al. 1997). PI 3-kinase has been shown to play a role both for growth factor-stimulated cell migration (Kundra et al. 1994; Wennstrom et al. 1994), and for integrin $\alpha6\beta4$-stimulated carcinoma cell invasion (Shaw et al. 1997) and FAK-promoted CHO cell migration on FN (Reiske et al. 1999). This latter study also demonstrated a specific role for FAK/PI 3-kinase association in cell migration, as a FAK point mutant, which selectively disrupts its binding to p85 but not Src abolished FAK-promoted cell migration (Reiske et al. 1999). The possible roles of FAK/PI 3-kinase association in other cellular functions await further investigation.

3.3
FAK Interaction with p130cas

p130cas is the first protein shown to bind to the proline-rich region of FAK through its SH3 domain (Polte and Hanks 1995; Harte et al. 1996). It is an adapter molecule containing one SH3 domain and multiple tyrosine phosphorylation sites within motifs for binding SH2 domains of Crk and other signaling molecules (Sakai et al. 1994; Nojima et al. 1995; Petch et al. 1995). p130cas was identified as a major tyrosine phosphorylated protein in v-Src or v-Crk transformed cells (Kanner et al. 1990) and its association with v-Src or v-Crk has also been demonstrated (Sakai et al. 1994). Similarly to FAK, p130cas is expressed ubiquitously and has been shown to be localized to focal adhesions in many cell types (Petch et al. 1995; Harte et al. 1996; Nakamoto et al. 1997). Furthermore, both cell adhesion to ECM and v-Src transformation induced tyrosine phosphorylation of p130cas as well as FAK (Sakai et al. 1994; Nojima et al. 1995; Petch et al. 1995; Harte et al. 1996). Lastly, p130cas has been identified as a substrate of several protein tyrosine phosphatases, including PTP-PEST (Garton et al. 1996), the *Yersinia YopH* protein (Black and Bliska 1997), PTP1B (Liu et al. 1996), and PTEN (Tamura et al. 1998), all of which have also been demonstrated to negatively regulate tyrosine phosphorylation and activity of FAK (Liu et al. 1998; Angers-Laustau et al. 1999; Tamura et al. 1998; Persson et al. 1997). Taken together, these results strongly suggest that p130cas function in close association with FAK in integrin-mediated signal transduction.

FAK interaction with p130cas has been demonstrated to play a critical role in the regulation of cell migration using CHO cells overexpressing FAK or its mutants (Cary et al. 1998). These studies also suggested that p130cas functions in cell migration might involve tyrosine phosphorylation of p130cas by FAK-associated Src and subsequent recruitment of an SH2 domain-containing signaling molecule to p130cas. Consistent with this, overexpression of p130cas or its major target, Crk, has also been shown to stimulate COS cell migration, which is dependent on their respective binding sites (Klemke et al. 1998). The mechanisms downstream of the p130cas/Crk complex in regulation of cell migration are not clear yet, although recent studies suggested several possibilities. The recently described Crk-binding protein DOCK180 has been suggested to regulate actin cytoskeleton dynamics (Hasegawa et al. 1996; Kiyokawa et al. 1998). Interestingly, tyrosine phosphorylation of p130cas in both cell adhesion and transformation correlated with p130cas translocation to cytoskeleton fractions (Polte and Hanks 1997). Therefore, the p130cas/Crk complex may control cell motility by regulation of the cytoskeleton. The p130cas/Crk complex has also been shown to activate both Erk (Schlaepfer et al. 1997) and JNK (Dolfi et al. 1998) signaling pathways, which could modulate expression of a variety of genes through phosphorylation and activation of transcription factors. It will be very interesting to investigate whether new gene expressions stimulated by FAK/p130cas complex formation could play a role in regulation of cell migration.

4
Biological Functions of FAK in Signaling by ECM

It is clear now that upon activation by integrins, FAK relays biochemical signals to several downstream signaling cascades by interacting with a number of important signaling molecules as discussed above. Activation of these signaling pathways contributes to a variety of biological functions regulated by integrin-mediated cell adhesion to ECM such as cell adhesion and spreading, cell migration, cell survival and apoptosis, and cell cycle regulation and proliferation (see Fig. 2).

4.1
Cell Adhesion and Spreading

A role of FAK in integrin-mediated cell adhesion and spreading and focal adhesion assembly was proposed when FAK was identified initially. This hypothesis was based on the observation that phosphorylation of FAK preceded stable

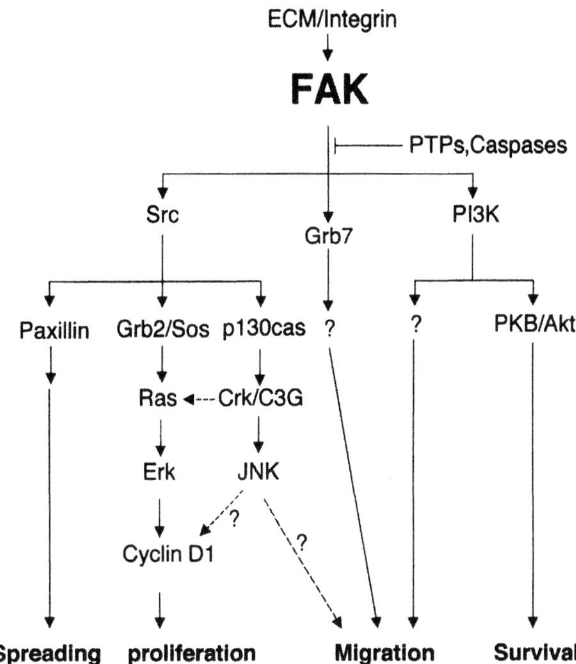

Fig. 2. FAK-mediated integrin signaling pathways. ECM binding to integrins leads to FAK activation. FAK relays integrin signals by binding to key signaling proteins such as Src and PI 3-kinase (*PI3K*). *Src/FAK* association is most likely responsible for cell spreading by phosphorylation of paxillin, for cell migration by activation of the *p130cas/Crk* pathway, for cell cycle progression and proliferation by activation of the *Grb2/Sos/Ras/Erk* pathway. *PI3K/FAK* association is responsible for cell migration by an unknown mechanism and possibly is also involved in regulation of cell survival through activation of *PKB/Akt*. Most recently, the adapter protein *Grb7* has been found to be involved in FAK-promoted cell migration by an unknown mechanism. Finally, protein tyrosine phosphatases (*PTPs*) and apoptosis-specific proteases (caspases) are most likely involved in downregulation of integrin/FAK signaling pathways

cell adhesion and cell spreading (Guan et al. 1991; Burridge et al. 1992). Inhibition of FAK phosphorylation was also shown to result in reduced cell adhesion (Schwartz et al. 1995). Surprisingly, however, FAK–/– fibroblasts derived from FAK knockout mouse embryo showed no defects in cell adhesion, but exhibited an increased number of focal adhesions (Ilic et al. 1995). These results strongly suggest that FAK is not involved in integrin-mediated cell adhesion. Consistent with this, microinjetion of a dominant-negative FAK inhibitor, FRNK (for FAK-related C-terminal nonkinase), or overexpression of FAK in CHO cells showed no effect on cell adhesion (Gilmor and Romer 1996; Cary et al. 1996).

Although change in cell spreading was not observed in various cells overexpressing FAK or its mutants (Hildebrand et al. 1993; Cary et al. 1996), the FAK–/– cells did exhibit somewhat reduced spreading (Ilic et al. 1995). In addition, overexpression of FRNK caused delayed cell spreading of chicken embryonic fibroblasts on FN by decreasing tyrosine phosphorylation of FAK. The corresponding reduction of tyrosine phosphorylation of paxillin has been suggested to be responsible for the delayed cell spreading (Richardson and Parsons 1996; Richardson et al. 1997).

4.2
Cell Migration

FAK has been proposed to play a role in cell migration based on the early observations that endothelial cell migration into wounded monolayer correlated with increased tyrosine phosphorylation and kinase activity of FAK (Romer et al. 1994) and that the rapid migration of keratinocytes in epidermal wound healing coincides with FAK expression in the migrating cells (Gates et al. 1994). Indeed, FAK–/– fibroblasts derived from FAK-knockout mouse embryo showed a significant decrease in cell migration compared with the cells from wild-type mice. The defects in cell migration have been suggested to be responsible for the general mesodermal deficiency and embryonic lethal phenotype of the knockout mice (Ilic et al. 1995). Similarly, inhibition of FAK by microinjection of a FRNK-like FAK C-terminal recombinant protein caused decreased motility of both fibroblasts and endothelial cells (Gilmore and Romer 1996). Finally, overexpression of FAK promoted CHO cell migration on FN in a Boyden Chamber assay system (Cary et al. 1996), which depended on FAK association with both SH2 domain-containing proteins Src, Fyn, and PI 3-kinase, as well as p130cas (Cary et al. 1996, 1998; Reiske et al. 1999).

The mechanism by which FAK promotes cell migration is not totally clear at present. Whereas FAK/Src association has been demonstrated to play a role in activation of Ras/Erk signaling pathways in several cell experimental systems (Buday and Downward 1993; Chardin et al. 1993; Egan et al. 1993; Zhao et al. 1998), Erk activation was not detected in CHO cells overexpressing FAK and with increased cell migration (Cary et al. 1996). In accordance with this, an inhibitor of Erk signaling pathways did not affect FAK-stimulated CHO cell migration and the FAK mutant with disrupted Grb2 binding site promoted CHO cell migration as much as the wild-type FAK (Cary et al. 1998). The formation of the p130cas/Crk complex stimulated by FAK has been shown to play an important role in promoting cell migration (Klemke et al. 1998). Recently, the p130cas/Crk complex has also been shown to activate the JNK signaling pathways (Tanaka et al. 1997). However, there is little support for a role of JNK in cell migration at present. Besides the FAK/Src/p130cas complex, recent mutational studies using a FAK mutant which selectively disrupted PI 3-kinase but not Src binding to the Y397 demonstrated a role for PI 3-kinase independent of Src or p130cas binding to FAK (Reiske et al. 1999). Finally, the adapter protein Grb7 has recently been shown to bind phosphorylated Y397 of FAK and functions to mediate FAK-stimulated cell migration independent of both PI 3-kinase and Src binding to the same site (Han and Guan 1999). Future challenges will be to determine how these independent downstream pathways can cooperate in the regulation of cell migration.

4.3
Cell Survival and Apoptosis

Cell adhesion to ECM has been shown to play a critical role in survival of many types of cells (Meredith et al. 1993; Ruoslahti and Reed 1994; Boudreau et al. 1995). Several recent studies have suggested that FAK signaling pathways in cell adhesion protect cells from apoptosis induced by cell detachment from ECM as well as other apoptosis-inducing agents. Overexpression of a membrane-anchored, constitutively activated FAK prevented the epithelial MDCK cells from anoikis (Chan et al. 1994; Frisch et al. 1996a). Conversely, inhibition of FAK either by FAK antisense oligonucleotides in tumor cell lines (Xu et al. 1996) or by microinjection of chicken embryonic fibroblasts with anti-FAK or the FRNK-like FAK C-terminal recombinant protein (Hungerford et al. 1996) induced apoptosis. In addition, several cell types induced to undergo apoptosis have shown concomitant FAK cleavage by specific caspases at specific sites on FAK (Croutch et al. 1996; Wen et al. 1997; Gervais et al. 1998). Besides decreasing the full-length functional FAK, FAK cleavage also generates an FRNK-like product that may further inhibit the function of any remaining FAK (Levkau et al. 1998).

The kinase activity and phosphorylation of Y397 of FAK have been shown to be required for FAK to prevent MDCK cells from anoikis (Frisch et al. 1996a), suggesting a key role for FAK association with Src and/or PI 3-kinase. However, expression of the constitutively activated FAK in these cells did not change the activity of Erk, indicating that the Erk signaling pathway is unlikely to be involved in FAK-regulated MDCK cell survival (Frisch et al. 1996b). Recently, substantial evidence has demonstrated that PI 3-kinase/Akt pathway serves as a survival signal in many cell types, including protecting epithelial cells from anoikis (Dudek et al. 1997; Philpott et al. 1997; Songyang et al. 1997; Kennedy et al. 1997; Kauffman-Zeh et al. 1997). Indeed, both PI 3-kinase and PKB/Akt were activated upon adhesion of epithelial cells to ECM (Khwaja et al. 1997; King et al. 1997). Together, these results suggest that PI 3-kinase/Akt pathway may play a critical role in FAK-regulated cell survival. Finally, a recent report also suggested that FAK prevented apoptosis through a pathway involving phospholipase A2, PKC, and p53, although the pathways from FAK to these proteins are not clear (Ilic et al. 1998).

4.4
Cell Cycle Progression and Proliferation

It has been known since the 1960s that cell adhesion to ECM is required for cell proliferation (MacPherson and Montagnier 1964; Freedman and Shin 1974). Most normal cells cannot grow without the anchorage to ECM even in the presence of sufficient growth factors and other nutrients. One of the hallmarks of cancer cells is their ability to proliferate independently of the anchorage. Recent studies have demonstrated that integrin-mediated cell adhesion to ECM controls the cell cycle progression through G1 phase by regulation of several G1 components of cell cycle machinery, including upregulation of cyclins D1, A, and E, upregulation of Cdk (for cyclin-dependent kinase) 2, 4, and 6, and downregulation of Cdk inhibitors p21 and p27 (Fang et al. 1996; Zhu et al. 1996; Assoian 1997).

Several recent studies have demonstrated a role for FAK in mediating integrin-regulated cell cycle progression and proliferation. Microinjection of the FRNK-like FAK C-terminal recombinant protein resulted in reduced DNA synthesis and cell cycle arrest in HUVEC and Balb/c 3T3 cells (Gilmore and Romer 1996). Conversely, overexpression of the membrane-anchored, constitutively activated FAK induced anchorage-independent growth of MDCK cells (Frisch et al. 1996a). Consistent with this, constitutive phosphorylation and activation of FAK has been observed in the anchorage-independent, v-Src transformed cells (Guan and Shalloway 1992). Interestingly, FAK−/− cells do not demonstrate reduced proliferation (Ilic et al. 1995), which could be a result of compensation by the elevated expression of FAK-related protein Pyk2 in these cells (Sieg et al. 1998). Alternatively, this may be due to the expression of mutant p53 (Furuta et al. 1995) or the polyoma middle T antigen (Ilic et al. 1998) in these cells, which could lead to a bypass of requirement of FAK in cell cycle regulation.

Very recent studies using inducible expression of FAK and mutants from our laboratory have provided direct evidence for FAK as a mediator of cell cycle regulation by integrins as well as some insight into the mechanisms (Zhao et al. 1998). While wild-type FAK increases DNA synthesis and accelerates the G1/S transition, a dominant-negative FAK construct inhibits these steps, and these effects correlate with expression of cyclin D1 and the cdk inhibitor p21. Furthermore, inhibition of cell cycle progression by the dominant negative FAK mutant required the Y397, suggesting a role for Src and/or PI 3-kinase binding to this site. These studies also correlated Erk activation and cell cycle progression in cells with induced expression of FAK or its mutants. These results suggested that FAK/Src association and its subsequent activation of Erk signaling pathway is responsible, at least partially, for integrin-regulated cell cycle progression. In addition, the required FAK Y397 also serves as binding site for PI 3-kinase, which is implicated in regulation of cell cycle progression and proliferation (Chou and Blenis 1995; Brennan et al. 1997). Therefore it will be interesting to examine whether the FAK/PI 3-kinase pathway is also involved in cell cycle progression in cell adhesion.

Acknowledgments. Research in the authors' laboratory is supported by NIH grants R01GM48050 and R01GM52890. Jun-Lin Guan is an Established Investigator of the American Heart Association. We thank Dong Cho Han, Hiroki Ueda, Heinz R. Reiske and Tanglong Shen for their critical reading of the manuscript and helpful suggestions.

References

Akiyama SK, Yamada SS, Yamada KM, LaFlamme SE (1994) Transmembrane signal transduction by integrin cytoplasmic domains expressed in single-subunit chimeras. J Biol Chem 269:15961–15964

Andre E, Becker-Andre M (1993) Expression of an N-terminally truncated form of human focal adhesion kinase in brain. Biochem Biophys Res Commun 190:140–146

Angers-Laustau A, Cote JF, Charest A, Dowbenko D, Spencer S, Lasky LA, Tremblay ML (1999) Protein tyrosine phosphotase-PEST regulates focal adhesion disassembly, migration, and cytokinesis in fibroblasts. J Cell Biol 144:1019–1031

Assoian RK (1997) Anchorage-dependent cell cycle progression. J Cell Biol 136:1–4

Avraham S, London R, Fu Y, Ota S, Hiregowdara D, Li J, Jiang S, Pasztor LM, White RA, Groopman JE, Avraham H (1995) Identification and characterization of a novel related adhesion focal adhesion kinase (RAFTK) from megakaryocytes and brain. J Boil Chem 270:27742–27751

Birge RB, Fajardo JE, Reichman C, Shoeson SE, Songyang Z, Cantley LC, Hanafusa H (1993) Identification and characterization of a high-affinity interaction between v-Crk and tyrosine phosphorylated paxillin in CT 10-transformed fibroblasts. Mol Cell Biol 13:4648–4656

Black DS, Bliska JB (1997) Identification of p130Cas as a substrate of *Yersinia YopH* (Yop51), a bacterial protein tyrosine phosphatase that translocates into mammalian cells and targets focal adhesions. EMBO J 16:2730–2744

Bockholt SM, Burridge K (1993) Cell spreading on extracellular matrix proteins induces tyrosine phosphorylation of tensin. J Biol Chem 268:14565–14567

Bottazzi ME, Assoian RK (1997) The extracellular matrix and mitogenic growth factors control G1 phase cyclins and cyclin-dependent kinase inhibitors. Trends Cell Biol 7:348–352

Boudreau N, Sympson CJ, Werb Z, Bissell MJ (1995) Suppression of ICE and apoptosis in mammary epithelial cells by extracellular matrix. Science 267:891–893

Brennan P, Babbage JW, Burgering BM, Groner B, Reif K, Cantrell DA (1997) Phosphatidylinositol 3-kinase couples the interleukin-2 receptor to the cell cycle regulator E2F. Immunity 7:679–689

Buday L, Downward J (1993) Epidermal growth factor regulates p21ras through the formation of a complex of receptor, Grb2 adapter protein, and Sos nucleotide exchange factor. Cell 73:611–620

Burgaya F, Girault JA (1996) Cloning of focal adhesion kinase, pp125FAK, from rat brain reveals multiple transcripts with different patterns of expression. Mol Brain Res 37:63–73

Burgaya F, Toutant M, Studler J-M, Costa A, Le Bert M, Gelman M, Girault J-A (1997) Alternatively spliced focal adhesion kinase in rat brain with increased autophosphorylation activity. J Biol Chem 272:28720–28725

Burguring BM, Coffer PJ (1995) Protein kinase B (c-Akt) in phosphatidylinositol-3-OH kinase signal transduction. Nature 376:599–602

Burridge K, Turner CE, Romer LH (1992) Tyrosine phosphorylation of paxillin and pp125FAK accompanies cell adhesion to extracellular matrix: a role in cytoskeletal assembly. J Cell Biol 119:893–903

Calalb MB, Polte TR, Hanks SK (1995) Tyrosine phosphorylation of focal adhesion kinase at sites in the catalytic domain regulates kinase activity: a role for Src family kinases. Mol Cell Biol 15:954–963

Calalb MB, Zhang X, Polte TR, Hanks SK (1996) Focal adhesion kinase tyrosine-861 is a major site of phosphorylation by Src. Biochem Biophys Res Commun 228:662–668

Cary LA, Chang JF, Guan J-L (1996) Stimulation of cell migration by overexpression of focal adhesion kinase and its association with Src and Fyn. J Cell Sci 109:1787–1794

Cary LA, Han DC, Polte TR, Hanks SK, Guan J-L (1998) Identification of p130Cas as a mediator of focal adhesion kinase-promoted cell migration. J Cell Biol 140:211–221

Chan P-Y, Kanner SB, Whitney G, Aruffo A (1994) A transmembrane-anchored chimerical focal adhesion kinase is constitutively activated and phosphorylated at tyrosine residues identical to pp125FAK. J Biol Chem 269:20567–20574

Chardin P, Camonis J, Gale WL, Van-Aelst L, Schlessinger J, Weigler MH, Bar-Sagi D (1993) Human Sos1, a guanine nucleotide exchange factor for Ras that binds to Grb2. Science 260:1338–1343

Chen H-C, Guan J-L (1994) Association of focal adhesion kinase with its potential substrate phosphatidylinositol 3-kinase. Proc Natl Acad Sci USA 91:10148–10152

Chen H-C, Appeddu PA, Parsons JT, Hildebrand JD, Schaller MD, Guan J-L (1995) Interaction of focal adhesion kinase with cytoskeletal protein talin. J Biol Chem 270:16995–16999

Chen H-C, Appenddu PA, Isoda H, Guan J-L (1996a) Phosphorylation of tyrosine 397 in focal adhesion kinase is required for binding phosphatidylinositol-3 kinase. J Biol Chem 271:26329–26334

Chen Q, Lin TH, Der CJ, Juliano RL (1996b) Integrin-mediated activation of MEK and mitogen-activated protein kinase is independent of Ras. J Biol Chem 271:18122–18127

Chou MM, Blenis J (1995) The 70-kDa S6 kinase: regulation of a kinase with multiple roles in mitogenic signaling. Curr Opin Cell Biol 7:806–814

Clark EA, Brugge JS (1995) Integrins and signal transduction pathways, the road taken. Science 286:233–235

Clark EA, Hynes RO (1996) Ras activation is necessary for integrin-mediated activation of extracellular signal-regulated kinase 2 and cytosolic phospholipase A2 but not for cytoskeletal organization. J Biol Chem 271:14814–14818

Cobb BS, Schaller MD, Leu T-H, Parsons JT (1994) Stable association of pp60src and pp59fyn with the focal adhesion-associated protein kinase, pp125FAK. Mol Cell Biol 14:147–155

Cooper JA, Howell B (1993) The when and how of Src regulation. Cell 73:1051–1054

Crouch DH, Fincham VJ, Frame MC (1996) Targeted proteolysis of the focal adhesion kinase pp125FAK during c-Myc-induced apoptosis is suppressed by integrin signalling. Oncogene 12:2689–2696

Derkinderen P, Toutant M, Burgaya F, Le Bert M, Siciliano JC, de Franciscis V, Gelman M, Girault J-A (1996) Regulation of a neuronal form of focal adhesion kinase by anandamide. Science 273:1719–1722

Dolfi F, Garcia-Guzman M, Ojaniemi M, Nakamura H, Matsuda M, Vuori K (1998) The adaptor protein Crk connects multiple cellular stimuli to the JNK signaling pathway. Proc Natl Acad Sci USA 95:15394–15399

Dudek H, Datta SR, Franke TF, Burnbaum MJ, Yao RJ, Cooper GM, Segal RA, Kaplan DR, Greenberg ME (1997) Regulation of neuronal survival by the serine-threonine protein kinase Akt. Science 275:661–665

Egan SE, Giddings BW, Brooks MW, Buday L, Sizeland AM, Weinberg R (1993) Association of Ras exchange protein Sos with Grb2 is implicated in tyrosine kinase, signal transduction and transformation. Nature 363:45–51

Eide BL, Turck CW, Escobedo JA (1995) Identification of Tyr-397 as the primary site of tyrosine phosphorylation and pp60src association in the focal adhesion kinase, pp125FAK. Mol Cell Biol 15:2819–2827

Fang F, Orend G, Watanabe N, Hunter T, Ruoslahti E (1996) Dependence of cyclin E-cdk2 kinase activity on cell onchorage. Science 271:499–502

Freedman VH, Shin S-I (1974) Cellular tumorigenicity in nude mice: correlation with cell growth in semi-solid medium. Cell 3:355–359

Frisch SM, Vuori K, Ruoslahti E, Chan-Hui, PY (1996a) Control of adhesion-dependent cell survival by focal adhesion kinase. J Cell Biol 134:793–799

Frisch SM, Vuori K, Ruoslahti E, Sicks S (1996b) A role for Jun-N-terminal kinase in anoikis; suppresson by Bcl-2 and crmA. J Cell Biol 135:1377–1382

Garton AJ, Flint AJ, Tonks NK (1996) Identification of p130Cas as a substrate for the cytosolic protein tyrosine phosphatase PTP-PEST. Mol Cell Biol 16:6408–6418

Gates RE, King LE Jr., Hanks SK, Manney LB (1994) Potential role for focal adhesion kinase in migrating and proliferating keratinocytes near eoidermal wounds and in culture. Cell Growth Differ 5:891–899

Gervais FG, Thornberry NA, Ruffolo SC, Nicholson DW, Roy S (1998) Caspases cleave focal adhesion kinase during apoptosis to generate a FRNK-like polypeptide. J Biol Chem 273:17102–17108

Gilmore AP, Romer LH (1996) Inhibition of focal adhesion kinase (FAK) signaling in focal adhesions decreases cell motility and proliferation. Mol Biol Cell 7:1209–1224

Guan J-L (1997) Role of focal adhesion kinase in integrin signaling. Int J Biochem Cell Biol 29:1085–1096

Guan J-L, Shalloway D (1992) Regulation of focal adhesion-associated protein tyrosine kinase by both cellular adhesion and oncogenic transformation. Nature 358:690–692

Guan J-L, Trevithick JE, Hynes RO (1991) Fibronectin/integrin interaction induces tyrosine phosphorylation of a 120-kDa protein. Cell Regul 2:951–964

Guinebault C, Payrastre B, Racaud-Sultan C, Mazarguil H, Breton M, Mauco G, Plantavid M, Chap H (1995) Integrin dependent translocation of phosphoinositide 3-kinase to the cytoskeleton of thrombin-activated platelets involves specific interactions of p85alpha with actin filaments and focal adhesion kinase. J Cell Biol 129:831–842

Han DC, Guan J-L (1999) Association of focal adhesion kinase with Grb7 and its role in migration J Biol Chem 274:24425–24430

Hanks SK, Calalb MB, Harper MC, Patel SK (1992) Focal adhesion protein tyrosine kinase on fibronectin. Proc Natl Acad Sci USA 89:8487–8491

Harte MT, Hildebrand JD, Burnham MR, Bouton AH, Parsons JT (1996) p130Cas, a substrate associated with v-Src and v-Crk, localizes to focal adhesions and binds to focal adhesion kinase. J Biol Chem 271:13649–13655

Hasegawa H, Kiyokawa E, Tanaka S, Nagashima K, Gotoh N, Shibuya M, Kurata N, Mtsuda M (1996) DOCK180, a major Crk-binding protein, alters cell mophology upon translocation to the cell membrane. Mol Cell Biol 16:1770–1776

Hens MD, DeSimone DW (1995) Molecular analysis and developmental expression of the focal adhesion kinase pp125FAK in *Xenopus laevis*. Dev Biol 170:274–288

Herzog H, Hicholl J, Hort YT, Sutherland JR, Shine J (1996) Molecular cloning and assignment of FAK2, a novel human focal adhesion kinase, to 8p11.2–p22 by nonisotopic in situ hybridization. Genomics 32:484–486

Hildebrand JD, Schaller MD, Parsons JT (1993) Identification of sequences required for the efficient localization of the focal adhesion kinase, pp127FAK, to cellular focal adhesion. J Cell Biol 123:993–1005

Hildebrand JD, Schaller MD, Parsons JT (1995) Paxillin, a tyrosine phosphorylated focal adhesion associated protein binds to the carboxyl terminal domain of focal adhesion kinase. Mol Biol Cell 6:637–647

Hildebrand JD, Taylor JM, Parsons JT (1996) An SH3 domain-containing GTPase-activating protein for Rho and Cdc42 associates with focal adhesion kinase. Mol Cell Biol 16:3169–3178

Horwitz AF, Duggan K, Buck C, Beekerle MC, Burridge K (1986) Interaction of plasma membrane fibronectin receptor with talin: a transmembrane linkage. Nature 320:531–533

Hungerford JE, Compton MT, Matter ML, Hoffstrom BG, Otey CA (1996) Inhibition of pp125FAK in cultured fibroblasts results in apoptosis. J Cell Biol 135:1383–1390

Hynes RO (1992) Integrins: versatility, modulation and signaling in cell adhesion. Cell 69:11–25

Ilic D, Furuta Y, Kanazawa S, Takeda N, Sobue K, Nakatsuji N, Nomura S, Fujimoto J, Okada M, Yamamoto T, Aizawa S (1995) Reduced cell motility and enhanced focal contact formation in cells from FAK-deficient mice. Nature 377:539–544

Ilic D, Almeida EAC, Schlaepfer DD, Dazin P, Aizawa S, Damsky CH (1998) Extracellular matrix survival signals transduced by focal adhesion kinase suppress p53-mediated apoptosis. J Cell Biol 143:547–560

Ishino M, Ohba T, Sasaki H, Sasaki T (1995) Molecular cloning of a cDNA encoding a phospho-protein, Efs, which contains a Src homology 3 domain and associates with Fyn. Oncogene 11:2331–2338

Juliano RL, Haskill S (1993) Signal transduction from the extracelllular matrix. J Cell Biol 120:577–585

Kanner SB, Reynolds AB, Vines RR, Parsons JT (1990) Monoclonal antibodies to individual tyro-sine-phosphorylated protein substrates of oncogene-encoded tyrosine kinase. Proc Natl Acad Sci USA 87:3328–3332

Kapron-Bras C, Fitz-Gibbon L, Jeevaratnam P, Wilkins J, Dedhar S (1993) Stimulation of tyrosine phosphorylation and accumulation of GTP-bound p21ras upon antibody-mediated α2β1 inte-grin activation in T-lymphoblastic cells. J Biol Chem 268:20701–20704

Kauffman-Zeh A, Rodrigues-Viciana P, Ulrich E, Gilbert C, Coffer P, Downward J, Evan G (1997) Suppression of c-Myc-induced apoptosis by Ras signaling through PI3K and PKB. Nature 385:544–548

Kennedy SG, Wagner AJ, Conzen SD, Jordan J, Bellacosa A, Tsichlis PN, Hay N (1997) The PI3K/Akt signaling pathway delivers an anti-apoptoic signal. Genes Dev 11:701–713

Khwaja A, Rodrigues-Viciana P, Wennstrom S, Warne PH, Downward J (1997) Matrix adhesion and Ras transformation both activate a phosphatidylinositol-3-OH kinase and protein kinase B/Akt cellular survival pathway. EMBO J 16:2783–2793

Kinashi T, Escobedo JA, Williams LT, Takatsu K, Springer TA (1995) Receptor tyrosine kinase stim-ulates cell-matrix adhesion by phosphatidylinositol-3 kinase and phospholipase C-gamma 1 pathways. Blood 86:2086–2090

King WJ, Mattaliano MD, Chan TO, Tsichlis PN, Brugge JS (1997) Phosphatidylinositol 3-kinase is required for integrin-stimulated Akt and Raf-1/mitogen-activated protein kinase pathway activation. Mol Cell Biol 17:4406–4418

Kiyokawa E, Hashimoto Y, Kurata T, Sugimura H, Matsuda M (1998) Evidence that DOCK180 up-regulates signals from the CrkII-p130cas complex. J Biol Chem 273:24479–24484

Klemke RL, Leng J, Molander R, Brooks PC, Vuori K, Cheresh DA (1998) CAS/Crk coupling serves as a "molecular switch" for induction of cell migration. J Cell Biol 140:961–972

Kornberg L, Earp SE, Turner CE, Procktop C, Juliano RL (1991) Signal transduction by integrins: increased protein tyrosine phosphorylation caused by clustering of β1 integrins. Proc Natl Acad Sci USA 88:8392–8396

Kornberg L, Earp HS, Parsons JT, Schaller M, Juliano RL (1992) Cell adhesion or integrin clus-tering increase phosphorylation of a focal adhesion-associated tyrosine kinase. J Biol Chem 267:23439–23442

Kulic G, Klippel A, Webber WJ (1997) Anti-apoptosis signaling by the IGF-I receptor, PI3K and Akt. Mol Cell Biol 17:1595–1606

Kundra V, Escobedo JA, Kazlauskas A, Kim HK, Rhee SG, Williams LT, Zetter BR (1994) regula-tion of chemotaxis by the platelet-derived growth factor-beta. Nature 367:474–476

Lauffenburger DA, Horwitz AF (1996) Cell migration: a physically integrated molecular process Cell 84:359–369

Lev S, Moreno H, Martinez R, Canoll P, Peles E, Musacchio JM, Plowman GD, Rudy B, Schlessinger J (1995) Protein tyrosine kinase PYK2 involved in Ca^{2+}-induced regulation of ion channel and MAP kinase functions. Nature 376:737–745

Levkau B, Herren B, Koyama H, Ross R, Raines EW (1998) Caspase-mediated cleavage of focal adhesion kinase pp125FAK and disassembly of focal adhesions in human endothelial cell apoptosis. J Exp Med 187:579–586

Lewis JM, Schwartz MA (1995) Mapping *in vivo* association of cytoplasmic proteins with inte-grin β1 cytoplasmic domain mutants. Mol Biol Cell 6:151–160

Liu F, Hill DE, Chernoff J (1996) Direct binding of the proline-rich region of protein tyrosine phosphatase 1B to the src homology 3 domain of p130Cas. J Biol Chem 271:31290–31295

Liu F, Sells MA, Chernoff J (1998) Protein tyrosine phosphatase 1B negatively regulates integrin signaling. Curr Biol 8:173–176

Lukashev ME, Sheppard D, Pytela R (1994) Disruption of integrin function and induction of tyrosine phosphorylation by the autonomously expressed $\beta1$ integrin cytoplasmic domain. J Biol Chem 269:18311–18314

MacPherson I, Montagnier L (1964) Agar suspension culture for the selective assay of cells transformed by polyoma virus. Virology 23:291–294

Matsuda M, Hashimoto Y, Muroya K, Hasegawa H, Kurata T, Tanaka S, Nakamura S, Hattori S (1994) Crk protein binds to two guanine nucleotide-releasing proteins for the Ras family and modulates nerve growth factor-induced activation of ras in PC12 cells. Mol Cell Biol 14:5495–5500

Meredith JE, Fazeli B, Schwartz MA (1993) The extracellular matrix as a cell survival factor. Mol Biol Cell 4:953–961

Miyamoto C, Akiyama S, Yamada KM (1995) Synergistic roles for receptor occupancy and Aggregation in integrin transmembrane function. Science 267:883–885

Nakamoto T, Sakai R, Honda H, Ogawa S, Ueno H, Suzuki T, Aizawa S, Yazaki Y, Hirai H (1997) Requirements for localization of p130Cas to focal adhesions. Mol Cell Biol 17:3884–3897

Nojima Y, Morino N, Mimura T, Hamasaki K, Furuya H, Sakai R, Sato T, Tachibana K, Morimoto C, Yazaki Y, Hirai H (1995) Integrin-mediated cell adhesion promotes tyrosine phosphorylation of p130Cas, a Src homology 3-containing molecule having multiple Src homology 2-binding motifs. J Biol Chem 270:15398–15402

Parsons JT (1996) Integrin-mediated signalling: regulation by protein tyrosine kinases and small GTP-binding proteins. Curr Opin Cell Biol 8:146–152

Persson C, Carballeira N, Wolf-Watz H, Fallman M (1997) The PTPase YopH inhibits uptake of *Yersinia*, tyrosine phosphorylation of p130Cas and FAK, and the associated accumulation of these proteins in peripheral focal adhesions. EMBO J 16:2307–2318

Petch LA, Bockholt SM, Bouton A, Parsons JT, Burridge K (1995) Adhesion-induced tyrosine phosphorylation of the p130 SRC substrate. J Cell Sci 108:1371–1379

Philpott KL, McCarthy MJ, Klippel A, Rubin LL (1997) Activated phosphatidylinositol-3 kinase and Akt kinase promote survival of superior cervical neurons. J Cell Biol 139:809–815

Polte TR, Hanks SK (1995) Interaction between focal adhesion kinase and Crk-associated tyrosine kinase substrate p130Cas. Proc Natl Acad Sci USA 92:10678–10682

Polte TR, Hanks SK (1997) Complexes of focal adhesion kinase (FAK) and Crk-associated substrate (p130^{Cas}) are elevated in cytoskeleton-associated fractions following adhesion and Src transformation. J Biol Chem 272:5501–5509

Reiske HR, Kao S-C, Cary LA, Guan J-L, Lai J-F, Chen H-C (1999) Requirement of phosphatidylinositol 3-kinase in focal adhesion kinase-promoted cell migration. J Biol Chem 274:12361–12366

Richardson A, Parsons JT (1995) Signal transduction through integrins: a central role for focal adhesion kinase? Bio Essays 17:229–236

Richardson A, Parsons JT (1996) A mechanism for regulation of the adhesion-associated protein tyrosine kinase pp125FAK. Nature 380:538–540

Richardson A, Malik RK, Hildebrand JD, Parsons JT (1997) Inhibition of cell spreading by expression of C-terminal domain of focal adhesion kinase (FAK) is rescued by coexpression of Src or catalytically inactive FAK: a role for paxillin tyrosine phosphorylation. Mol Cell Biol 17:6906–6914

Romer LH, Mclean N, Turner CE, Burridge K (1994) Tyrosine kinase activity, cytoskeleton organization, and motility in human vascular endothelial cells. Mol Biol Cell 5:349–361

Ruoslahti E, Reed JC (1994) Anchorage dependence, integrins, and apoptosis. Cell 77:477–478

Sabe H, Hata A, Okada M, Nakagawa H, Hanafusa H (1994) Analysis of the binding of the Src. homology 2 domain of Csk to tyrosine-phosphorylated proteins in the suppression and mitotic activation of c-Src. Proc Natl Acad Sci USA 91:3984–3988

Sakai R, Iwamatsu A, Hirano N, Ogawa S, Tanaka T, Mano H, Yazaki Y, Hirai H (1994) A novel signaling molecule, p130, forms stable complexes in vivo with v-Crk and v-Src in a tyrosine phosphorylation-dependent manner. EMBO J 13:3748–3756

Sasaki H, Nagura K, Ishino M, Tobioka H, Kotani K, Sasaki T (1995) Cloning and characterization of cell adhesion kinase β, a novel protein-tyrosine kinase of the focal adhesion kinase subfamily. J Biol Chem 270:21206–21219

Schaller MD, Parsons JT (1994) Focal adhesion kinase and associating proteins. Curr Opin Cell Biol 6:705–710

Schaller MD, Parsons JT (1995) pp125FAK-dependent tyrosine phosphorylation of paxillin creates a high affinity binding site for Crk. Mol Cell Biol 15:2635–2645

Schaller MD, Sasaki T (1997) Differential signaling by the focal adhesion kinase and cell adhesion kinase beta. J Biol Chem 272:25319–25325

Schaller MD, Borgman CA, Cobb BS, Reynolds AB, Parsons JT (1992) pp125FAK, a structurally distinctive protein tyrosine kinase associated with focal adhesions. Proc Natl Acad Sci USA 89:5192–5196

Schaller MD, Borgman CA, Parsons JT (1993) Autonomous expression of a noncatalytic domain of the focal adhesion-associated protein tyrosine kinase pp125FAK. Mol Cell Biol 13:785–791

Schaller MD, Hildebrand JD, Shannon JD, Fox JX, Vines RR, Parsons JT (1994) Autophosphorylation of the focal adhesion kinase, pp125FAK, directs SH2-dependent binding of pp60src. Mol Cell Biol 14:1680–1688

Schaller MD, Otey CA, Hildebrand JD, Parsons JT (1995) Focal adhesion kinase and paxillin bind to peptides mimicking beta integrin cytoplasmic domains. J Cell Biol 130:1181–1187

Schlaepfer DD, Hunter T (1996) Evidence for in vivo phosphorylation of the Grb2 SH2-domain binding site on focal adhesion kinase by Src-family protein-tyrosine kinases. Mol Cell Biol 16:5623–5633

Schlaepfer DD, Hanks SK, Hunter T, van der Geer P (1994) Integrin-mediated signal transduction linked to Ras pathway by Grb2 binding to focal adhesion kinase. Nature 372:786–791

Schlaepfer DD, Broome MA, Hunter T (1997) Fibronectin-stimulated signaling from a focal adhesion kinase-c-Src complex: involvement of the Grb2, p130Cas, and Nck adapter proteins Mol Cell Biol 17:1702–1713

Schwartz MA, Schaller MD, Ginsberg MH (1995) Integrins: emerging paradigms of signal transduction. Annu Rev Cell Dev Biol 11:549–599

Serve H, Yee NS, Stella G, Sepp-Lorenzino L, Tan JC, Besmer P (1995) Differential roles of PI 3 kinase and Kit tyrosine 821 in Kit receptor-mediated proliferation, survival and cell adhesion in mast cells. EMBO J 14:473–483

Shaw LM, Rabinovitz I, Wang HH, Toker A, Mercurio AM (1997) Activation of phosphoinositide 3-OH kinase by the α6β4 integrin promotes carcinoma invasion. Cell 91:949–960

Sieg DJ, Ilic D, Jones KC, Damsky H, Hunter T, Schlaepfer DD (1998) Pyk2 and Src-family protein-kinases compensate for the loss of FAK in fibronectin-stimulated signaling events but Pyk2 does not fully function to enhance FAK(−/−) cell migration. EMBO J 17:5933–5947

Sinnett-Smith J, Zachary I, Valverde AM, Rozengurt E (1993) Bombesin stimulation of p125 focal adhesion kinase tyrosine phosphorylation. J Biol Chem 268:14261–14268

Solowska J, Guan J-L, Marcantonio EE, Trevithick JE, Buck CA, Hynes OR (1989) Expression of normal and mutant avian integrin subunits in rodent cells. J Cell Biol 109:853–861

Songyang Z, Baltimore D, Cantley LC, Kaplan DR, Franke TF (1997) Interleukin 3-dependent survival by the Akt protein kinase. Proc Natl Acad Sci USA 94:11345–11350

Tachibana K, Sato T, D'Avirro N, Morimoto C (1995): Direct association of pp125FAK with paxillin, the focal adhesion-targeting mechanism of pp125FAK. J Exp Med 182:1089–1100

Tachibana K, Urano T, Fujita H, Ohashi Y, Kamiguchi K, Iwata S, Hirai H, Morimoto C (1997) Tyrosine phosphorylation of Crk-associated substrates by focal adhesion kinase. J Biol Chem 272:29083–29090

Tamura M, Gu J, Matsumoto K, Aota S, Parsons R, Yamada KM (1998) Inhibition of cell migration, spreading and focal adhesions by tumor suppressor PTEN. Science 280:1614–1617

Tanaka S, Morishita T, Hashimoto Y, Hattori S, Nakamura S, Shibuya M, Matsuoka K, Takenawa T, Kurata T, Nagashima K (1994) C3G, a guanine nucleotide-releasing protein expressed ubiquitously, binds to the Src homology 3 domains of Crk and Grb2/Ash proteins. Proc Natl Acad Sci USA 91:3443–3447

Tanaka S, Hanafusam H (1998) Guanine-nucleotide exchange protein C3G activates JNK1 by a Ras-independent mechanism. J Biol Chem 273:1281–1284

Tanaka S, Ouchi T, Hanafusa H (1997) Downstream of Crk adapter signaling pathway: activation of Jun kinase by v-Crk through the guanine nucleotide exchange protein C3G. Proc Natl Acad Sci USA 94:2356–2361

Thomas JW, Ellis B, Boerner RJ, Knight WB, White GC, Schaller MD (1998) SH2- and SH3 mediated interactions between focal adhesion kinase and Src. J Biol Chem 273:577–583

Turner CE, Miller J (1994) Primary sequence of paxillin contains putative SH2 and SH3 domain binding motif and multiple LIM domains: identification of a vinculin and p125FAK binding region. J Cell Sci 107:1583–1591

Turner CE, Schaller MD, Parsons JT (1993) Tyrosine phosphorylation of the focal adhesion kinase pp125FAK during development: relation to paxillin. J Cell Sci 105:637–645

Vuori K, Ruoslahti E (1995) Tyrosine phosphorylation of p130Cas and cortactin accompanies integrin-mediated cell adhesion to extracellular matrix. J Biol Chem 270:22259–22262

Vuori K, Hirai H, Aizawa S, Ruoslahti E (1996) Introduction of p130cas signaling complex formation upon integrin-mediated cell adhesion: a role for Src family kinases. Mol Cell Biol 16:2606–2613

Wary KK, Mainiero F, Isakoff SJ, Marcantonio EE, Giacotti FG (1996) The adapter protein Shc couples a class of integrins to the control of cell cycle progression. Cell 87:733–743

Weiner TM, Liu ET, Craven RJ, Cance WG (1993) Expression of focal adhesion kinase gene and invasive cancer. Lancet 342:1024–1025

Wen L-P, Fahrni JA, Troie S, Guan J-L, Orth K, Rosen GD (1997) Cleavage of focal adhesion kinase by caspases during apoptosis. J Biol Chem 272:26056–26061

Weng Q-P, Andrabi K, Klippel A, Kozlowski MT, Williams LT, Avruch J (1995) Phosphatidylinositol-3 kinase signals activation of p70 S6 kinase in situ through site-specific p70 phosphorylation. Proc Natl Acad Sci USA 92:5744–5748

Wennstrom S, Siegbahn A, Yokote K, Arvidsson AK, Heldin CH, Mori S, Claesson WL (1994) Membrane ruffling and chemotaxis transduced by the PDGF beta-receptor require the binding site for phosphatidylinositol-3' kinase. Oncogene 9:651–660

Whitney GS, Chan PY, Blake J, Cosand WL, Neubauer MG, Aruffo A, Kanner SB (1993) Human T and B lymphocytes express a structurally conserved focal adhesion kinase, pp125FAK. DNA Cell Biol 12:823–830

Xing Z, Chen H-C, Nowlen JK, Taylor SJ, Shalloway D, Guan J-L (1994) Direct interaction of v-Src with the focal adhesion kinase mediated by the Src SH2 domain. Mol Biol Cell 5:413–421

Xiong W, Parsons JT (1997) Induction of apoptosis after expression of PYK2, a tyrosine kinase structurally related to focal adhesion kinase. J Cell Biol 139:529–539

Xiong WC, Machlem M, Parsons JT (1998) Expression and characterization of splice variants of Pyk2, a focal adhesion kinase-related protein. J Cell Sci 273:15765–15772

Xu L-H, Owens LV, Sturge GC, Yang X, Liu ET, Craven RJ, Cance WG (1996) Attenuation of the expression of the focal adhesion kinase induces apoptosis in tumor cells. Cell Growth Differ 7:413–418

Zhang X, Wright CV, Hanks SK (1995) Cloning of a *Xenopus laevis* cDNA encoding focal adhesion kinase (FAK) and expression during early development. Gene 160:219–222

Zhao J-H, Reiske H, Guan J-L (1998) Regulation of the cell cycle by focal adhesion kinase. J Cell Biol 143:1997–2008

Zheng C, Xing Z, Bian ZC, Guo C, Akbay A, Warner L, Guan J-L (1998) Differential regulation of Pyk2 and focal adhesion kinase (FAK). The C-terminal domain of FAK confers response to cell adhesion. J Biol Chem 273:2384–2389

Zhu X, Ohtsubo M, Bohmer RM, Roberts JM, Assoian RK (1996) Adhesion-dependent cell cycle progression linkrd to the expression of cyclin D1, activation of cyclin E-cdk2, and phosphorylation of the retinoblastoma protein. J Cell Biol 133:391–403

Interaction Between Cells and Extracellular Matrix: Signaling by Integrins and the Elastin-Laminin Receptor

J. Labat-Robert[1] and L. Robert[1]

1
Introduction

Precise knowledge about extracellular matrix (ECM) components was very slow in emerging. In the middle of this century most text books mentioning connective tissues, such as histology books, mentioned only collagen and acid mucopolysaccharides. Proteins copurified with such polysaccharides were considered as impurities. Elastin was ignored as an artifact because of the harsh conditions used for its purification heating tissues to 100°C in 0.1 N NaOH! The importance of the mechanical properties of ECM components was progressively revealed by the crucial role of cartilage rich in ECM, in joint function, and its diseases as osteoarthritis. The isolation of intact proteoglycans by Hascall and Sajdera in the 1960s revealed the complexity of proteoglycan aggregates and showed the way to understanding them. Similarly, the demonstration of collagen type II in cartilage by Miller in Carl Piez's Laboratory at about the same time opened the way for the progressive identification of a score of different collagen types. The introduction of the methods of molecular biology and genetic engineering into matrix biology, a term coined by E.A. Balazs (1970), who did a great deal to rationalize the nomenclature of this field, accelerated the identification and sequencing of genes coding for ECM components otherwise difficult to isolate. Books and proceedings of colloquia, especially that of the first NATO meeting on ECM organized by Fitton-Jackson et al. in 1964 and the second meeting organized by E.A. Balazs in 1969 (Balazs 1970) give plenty of information on this heroic period of research in what is now a rapidly expanding field.

Even slower to emerge was the concept of close interactions between cells producing ECM components and the extracellular macromolecules that these cells synthesize, that is cell-matrix interactions. The most important impetus came from the work on fibronectin (for reviews see Hynes 1990; Mosher 1989). Described first as cold insoluble globulin, a plasma component coprecipitating with fibrinogen, it was soon identified in tissues and its role in cell matrix interactions explored. The great interest stimulated by this work on cell-matrix

[1] Laboratoire d'Ophtalmologie, Hôtel-Dieu, 1 Place du Parvis Notre-Dame, 75181 Paris cedex 04, France, and Centre de Recherche Bioclinique sur le Vieillissement, Groupe Hospitalier Charles Foix-Jean Rostand, 7 avenue de la République, 94205 Ivry sur Seine, France

Progress in Molecular and Subcellular Biology, Vol. 25
A. Macieira-Coelho (Ed.)
© Springer-Verlag Berlin Heidelberg 2000

interactions led rapidly (within about two decades) to a much better understanding of the dynamic nature of cell-matrix interactions. This new way of viewing matrix components stimulated work in this area of matrix-oriented cell biology (Hay 1991). Functional consequences of such interactions were explored, such as the decreased expression and degradation of fibronectin in malignant tissues (Labat-Robert et al. 1981a) and its progressive overexpression during aging (Labat-Robert et al. 1981b; Rasoamanantena et al. 1993). This reorientation of the way of thinking in matrix biology was probably the main stimulus for the isolation of cell receptors interacting with fibronectin and later with other matrix components (Pierschbacher and Ruoslathi 1984; Hynes 1987). As most of this work was carried out by biochemists not trained in pharmacology, the first descriptions of these new receptors, the integrins, insisted only on their role as mechanical links between cells and ECM components. The understanding of their role in cell-matrix communication, in message transmission between cells and matrix components has emerged only recently. Another curiosity of this early work on cell-matrix interactions was that no mention was made of cell-elastin interactions. This prompted us to look for cell receptors mediating the interaction of cells with elastic fibers. Thanks to the medical-pharmacological orientation of our laboratory it appeared immediately (during the early 1980s) of crucial importance to explore the dynamic, pharmacological aspects of these interactions. To the best of our knowledge, the first functional, physio-pharmacological studies performed on matrix receptors were those published from our laboratory on the elastin receptor (Fülöp et al. 1986; Hornebeck et al. 1986; Jacob et al. 1987; Robert and Hornebeck 1986; Varga et al. 1988, 1989). These results will be described in section 3. In recent years, several investigators have studied message transfer by integrins. This topic has become one of the most important subjects of matrix-oriented research at the end of this century.

2
Cell-Matrix Interactions Mediated by Integrins

2.1
Signaling by Integrins

Among cell membrane receptors, namely cadherins, selectins, members of the Ig superfamily, and integrins, when extracellular matrix is concerned integrins are the major class of receptors (Hynes 1987; Defilippi et al. 1997). They are heterodimers composed of noncovalently linked α and β subunits that can be alternatively spliced. Sixteen α-subunits can associate with eight β subunits giving about 22 heterodimers. Alternative forms show distinct and specific functions (Defilippi 1997; Fornaro and Languino 1997; DeMelker and Sonnenberg 1999). They comprise a large extracellular domain, a short transmembrane segment, and usually a short cytoplasmic domain linking extracellular matrices with the actin cytoskeleton, with one exception, the large β4 subunit cytoplasmic domain which links extracellular matrix and intermediate filaments. Some integrins can bind to the same ligand, others are specific to one ligand. Most cells express multiple receptors on their surface. Some are ubiquitous molecules expressed on a variety of cells, others are expressed only on a given cell type, for example, leukocytes. Considered at the beginning as recognition molecules, integrins are, in fact, capable of transducing outside-in or inside-out signals and play a role in cytoskeletal organization, cell cycle, cell survival, and cell differentiation. It has to be remembered that integrins may show signaling differences.

When the extracellular ligand-binding pocket of an integrin is occupied by a matrix ligand, this binding being affected by divalent cations, there is a clustering of the receptors in the plane of the cell membrane that is followed by changes in integrin conformation, modification of $[pH]_i$ due to activation of a Na^+-K^+ antiporter, increase in $[Ca]_i$, a reorganization of cytoskeletal proteins which could play a role in assembling the transduction machinery, activation of tyrosine and serine/threonine kinases, and tyrosine phosphorylation of protein kinases. The role of cytoskeletal proteins in signaling comes from the observation of the effect of cytochalasin D that causes actin depolymerization and interferes with integrin signaling. Cytoplasmic domains of integrins also play a role in regulating integrin-ligand binding. Two distinct regions differentially regulate this binding which can also be modulated by cytoplasmic proteins that interact with integrin cytoplasmic domains.

One of the early events following integrin stimulation is the phosphorylation of the FAK family. This is a common response induced by most integrins. ECM receptors have no enzymatic activity of their own; they have to be coupled to cytosolic protein tyrosine kinases to transduce signals. FAK is a cytoplasmic kinase that becomes activated via paxillin and talin, followed by a reorganization of the cytoskeleton.

FAK is also at the cross-road of several signaling pathways, such as the MAPK pathway. It undergoes multityrosine phosphorylations that create binding sites for SH2 domains of Src family kinases, for SH2 domain of Grb-2. Activation of the MAP-kinase pathway can modulate the transcriptional activity in the nucleus. MAP-kinases are serine/threonine kinases that mediate a number of signal transductions. When the residue Y 925 is phosphorylated, it can fix itself to the SH2 domain of the adaptor Grb-2 which, in its turn, will associate with the SOS exchange factor and activate Ras-GDP into Ras-GTP. These small G-proteins are molecular switches which oscillate between two conformations according to the linked nucleotide. When linked to GDP, they are inactive, with GTP attached they are activated. They form a complex with the GDI protein which acts as an inhibitor of the dissociation of GDP. The exchange factor then facilitates the replacement of GDP by GTP and thus activates the MAP-kinase pathway. GTP will then be hydrolyzed by the action of a GAP protein, a GTP-ase activating protein, liberating GDP. During the cascade, two kinases are activated by external factors: Erk-1 et Erk-2 → MAPKKKinase → MAPKkinase which, on its turn, activates MAPkinase, which after a double phosphorylation will translocate to the nucleus and activate immediate early genes such as Jun or Myc, controlling the cell division cycle. These factors interact with DNA and activate the synthesis of cyclins. Cyclins control the cell cycle by reacting with Cdk (cycline-dependent kinases) and serine/threonine kinases, assuming the control of the cell cycle in G1. Inhibition of FAK stops the cell cycle and induces apoptosis.

Besides the Ras family, the Rho family of small GTP-binding proteins is also required for signaling to the actin cytoskeleton induced by activation of integrins. Rho controls stress fiber assembly and the related proteins Rac regulate formation of lamellipodia and Cdc42 filipodia formation and this correlates with cell cycle progression. Cell survival and cell cycle progression are controlled by distinct pathways.

Integrins can also activate lipid kinases, phospholipases, and PKC. They also act synergistically with growth factors.

If cells are not attached to the extracellular matrix, most of them undergo apoptosis [Frisch and Francis 1994; (anoikis) Frisch and Ruoslahti 1997]. Among the pathways implicated in this protection, one involves activation of focal adhesion kinases, because their inhibition leads to apoptosis. FAK contains two proline-rich domains which can bind to SH3 domains as the regulatory domain p85 of PI3-kinase. Besides this regulatory domain, PI3-kinase contains a catalytic domain which allows the synthesis of phosphatidylinositide triphosphate, which recruits protein kinase Akt/PKB. After phosphorylation Akt/PKB phosphorylates a pro-apoptotic protein, Bad, of the Bcl-2 family which becomes complexed and cannot dimerize and induce apoptosis. This PI3 kinase-AKT-Bad connection is the major mechanism of apoptosis control. Another pathway is the bcl-2/caspase connection. Dissociation of cell-matrix interaction leads to caspase activation through JNK pathway and bcl-2 blocking.

Among extracellular matrix components which can be ligands of integrins, there are collagens, glycoproteins such as fibronectin, laminin, tenascin, fibrillin, thrombospondin, and vitronectin, and proteoglycans such as perlecan (Kühn, 1997). Among the recognition sites for integrins, the most frequent is the tripeptide Arg-Gly-Asp (RGD) (Pierschbacher and Ruoslahti 1984; Pfaff 1997). More than half the integrins bind to this site, the specificity being given by the structure of the RGD loop protruding from the body of the protein, the conformation of RGD tripeptide, and also by the amino acid sequences flanking this motif.

2.2
Evolution of Integrin Signaling with Aging

Cell-matrix interactions vary from embryogenesis to the end of life because the extracellular matrix composition is not constant during the whole life. It varies as a function of age, during development, and all through maturation and aging. It is also modified in pathological states. During our studies we could show, using cell cultures, explant cultures, and animals of different ages, that fibronectin biosynthesis increased as a function of age at both the mRNA and protein levels (Boyer et al. 1991; Rasoamanantena et al. 1993). This was also found in diseases mimicking an accelerated aging such as the Werner syndrome (Rasoamanantena et al. 1994). These quantitative alterations can be accompanied by qualitative changes, such as protein fragmentation leading to new properties exhibited by fibronectin fragments. Since senescent fibroblasts have a decreased proliferative capacity (Goldstein 1990) associated with high fibronectin levels, Hu et al. (1996) carried out experiments on the main fibronectin integrin, $\alpha5\beta1$. They could observe alterations at both the RNA and protein levels, in polypeptide localization, and in $\alpha5\beta1$-mediated cellular functions. They compared normal young cells, senescent normal cells and Werner syndrome cells and demonstrated reduced levels of the $\alpha5$-polypeptide on senescent and Werner syndrome cell membranes as well as a decreased accessibility of this integrin to monoclonal antibody, which could explain the weaker ligand binding. $\beta1$ polypeptide, on its turn, showed a slower rate, of processing which could lead to the formation of immature heterodimers with a reduced binding affinity to fibronectin. In senescent cells and in Werner syndrome cells there was also a decline in cell adhesion and proliferation. A defect in transmembrane signaling has been described also in dermal fibroblasts from Down's syndrome patients, when compared to normal fibroblasts (Flickinger et al. 1992).

3
Cell-Matrix Interactions Mediated by the Elastin-Laminin Receptor

3.1
Signaling by the Elastin-Laminin Receptor

As stated in the introduction, our search for an elastin-recognizing receptor started from the necessity of close coupling of vascular smooth muscle cells to elastic laminae in order to transduce contraction-relaxation cycles of cells to the matrix components of the vascular wall. Elastic laminae are major macromolecular components of elastic arteries (Fig. 1). They are synthesized by smooth muscle cells (SMCs) which also synthesize most other components of the vascular matrix. Our first observations concerned the in vitro reconstitution of cell-elastin interactions, as seen in tissues (Fig. 1). It could be shown that the coincubation of dermal fibroblasts and vascular SMCs with micronized, highly purified elastin fibers resulted in a slow but solid association of cells and fibrils (Hornebeck et al. 1986). The kinetics of this process could be followed by video recording over longer time periods (Groult et al. 1998). The morphometric evaluation of such video recordings revealed regularities in cell-elastin interactions (Groult et al. 1998) Video recording in inverted Rose chambers confirmed previous findings showing that cells adhere strongly

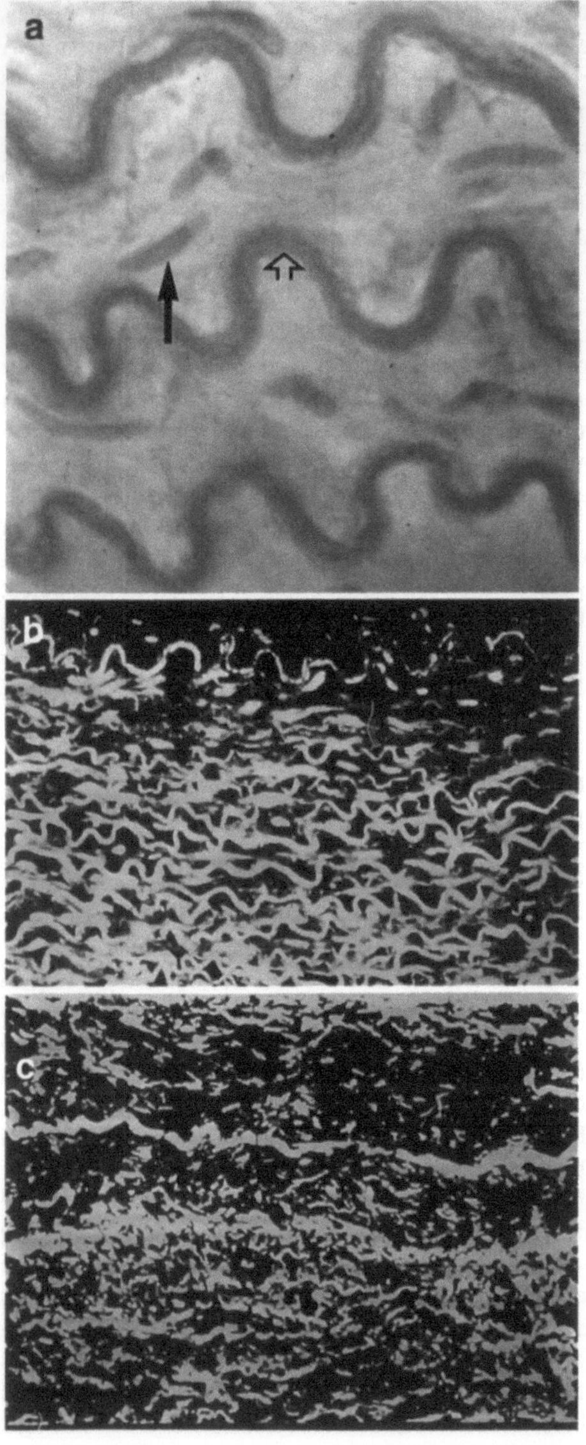

to elastin fiber. Elastase or trypsin, but not collagenase, could dissociate fibers from cells. In order to explore a possible competition between insoluble elastin fibers and soluble elastin peptides, K-elastin (KOH-degraded, high molecular weight elastin peptides: Jacob and Hornebeck 1985) were added to the cells before fibrous elastin. To our surprise, not only did these peptides not compete out fibers for cell adhesion but they strongly accelerated it (Hornebeck et al. 1986; Robert and Hornebeck 1986). This "induced" adhesion was attributed to a receptor-stimulated synthesis of an adhesive membrane complexe, elastonectin. Indeed, adhesion could be inhibited by cycloheximide and did not occur at low temperature (+4°C). Immunostaining and confocal microscopy confirmed the presence of membrane localized components mediating cell adhesion to elastin fibers (Perdomo et al. 1994). These observations prompted us to explore the transduction mechanism of this receptor. Gel-electrophoretic profiles suggested three main components of 51, 61, and 67 kDa, similar to those identified by Hinek and Mecham in their experiments (1990).

Only the ~67-kDa elastin-binding component of this receptor was characterized and suggested as being identical to the nonintegrin laminin receptor (Mecham et al. 1989).

In our experiments on vasoactivity, laminin acted as an antagonist of the receptor (Faury et al. 1995). Recent experiments by G. Faury (in prep.) concerning the action of laminin on the endothelium- and NO-mediated vasorelaxing action of elastin peptides confirmes and extends the inhibitory action of laminin as mentioned previously (Faury et al. 1995) and justify the (provisory) designation of this receptor as the elastin-laminin receptor (ELR). Work in R. Mecham's laboratory suggested an implication of the 67-kDa subunit of this receptor (or elastin-binding protein, EBP) in elastin fiber formation as a chaperone protein (Mecham 1991).

Using a monoclonal antibody, this 67-kDa subunit could be demonstrated on a variety of cell types such as endothelial cells, fibroblasts, SMCs, and also mononuclear leukocytes – monocytes, PMNs, and lymphocytes –.

Using ^{45}Ca or fluorescent Ca-probes, a steady influx of calcium in SMCs, fibroblasts, leukocytes, and endothelial cells could be demonstrated in cell culture conditions or in vitro experiments when elastin peptides were added (Fig. 2A). The efflux of Ca from preloaded cells was inhibited by elastin peptides indicating a calmodulin dependent mechanism (Fig. 2B; Jacob et al. 1987).

Pharmacological and signal transduction studies were largely facilitated by the use of mononuclear cells (Varga et al. 1988, 1989). In collaboration with Marie-Paule Jacob and Tamas Fülöp, it could be shown that in human

Fig. 1. a Elastic laminae in the aorta (↔) with smooth muscle cells (◆) which produce most of the extracellular matrix of the vessel wall. **b** Difference in the continuity of elastic laminae of aorta in young (above) and old (>60 years, below) individuals indicates a progressive elastolysis with age

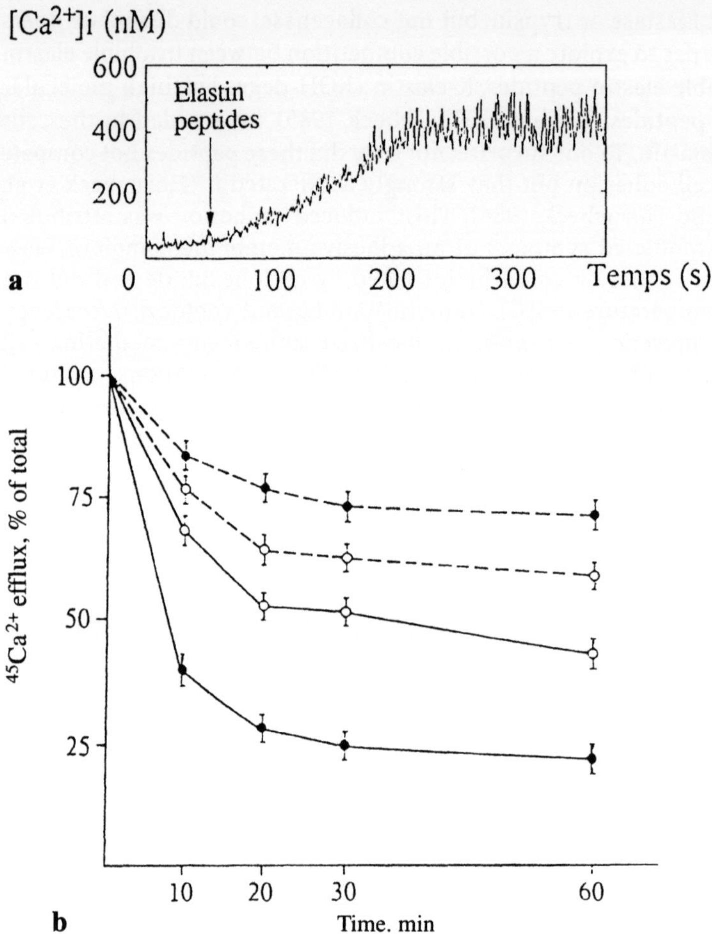

Fig. 2a,b. Ca influx in endothelial cells (**a**) and inhibition of Ca out flux from mononuclear cells (**b**), in presence of elastin peptides. **a.** The effect of elastin peptides (10^{-2} mg/ml) is recorded on the intracellular free Ca of cells (Faury et al. 1995). **b.** The uninhibited outflux of Ca from pre-loaded monocytes (-●-). The effect of elastin peptides (-O- $1 \mu g\,ml^{-1}$) is compared to that of trifluoperazin $1 \mu M$ (--●--) alone or with $1 \mu g\,ml^{-1}$ elastin peptides (--O--) (Faury et al. 1995; Jacob et al. 1987)

mononuclear cells the elastin receptor is linked to a pertussis toxin-sensitive G-protein which activates phospholipase C, resulting in a rapid mobilization of the phosphoinositol pathway, activation of phosphokinase C (PKC), production of diacylglycerol (DAG), and a rapid increase in intracellular free calcium (Fig. 2A). Another intriguing finding was the strong inhibition of ouabain-dependent K^+ influx in K elastin-stimulated cells (Varga et al. 1989; Fülöp et al. 1990).

Another important function of the elastin receptor in mononuclear cells was the rapid increase in O_2 consumption accompanied by the extrusion of elastase and cathepsin G from PMN leukocytes as well as the liberation of superoxide.

Increase of elastase-type endopeptidase production by activation of the receptor with K-elastin could be demonstrated with cultured fibroblasts and SMCs (Archilla-Marcos and Robert 1993).

Signal transduction appeared to vary with cell type. In vascular endothelial cells, G. Faury et al. carried out patch clamp experiments in order to characterize the calcium channels activated by the elastin receptor (Faury et al. 1998A). Using HUVECs, this technique indicated the opening of low-specificity calcium channels by interaction with the cytoskeleton-actin filaments and was independent in these cells of phosphoinositol metabolism. Both cytoplasmic and nuclear free calcium levels increased in presence of K-elastin added in concentrations found in circulating human serum (Fülöp et al. 1990; Bizbiz et al. 1997). This action could be inhibited by lactose, an antagonist of the 67-kDa subunit which has some properties of a galectin (Hinek and Mecham 1990).

Using synthetic peptides derived from elastin sequences, it could be shown that the shortest sequences still activating the receptor are VGV and PGV. The repeating hexapeptide, VGVAPG, present in seven copies in the human elastin gene and shown to be chemotactic to mononuclear cells (Senior et al. 1984) and also to SMCs (Ooyama et al. 1987) was also active (Faury et al. 1998B).

3.2
Effect of Age on the Elastin-Laminin Receptor

When mononuclear cells of young (30–40 years) and old (>65 years) individuals were studied, it could be shown that the calcium transients triggered by elastin peptides was lower in old cells as compared to young cells (Varga et al. 1990). Similar results were obtained with a different receptor, triggered by the bacterial peptide FMLP (Ghuysen-Itard et al. 1993). When PMN leukocytes were taken from geriatric patients, about one third of all investigated cases showed no return of intracellular calcium levels to prestimulation values (Ghuysen-Itard et al. 1993). Loss of calcium homeostasis appeared thus to characterize these aging cells. This effect was accompanied by a strong decrease in the mobilization of the inositol phosphate metabolism. Curiously, however, the liberation of superoxide production did not decrease, but increased somewhat in old elastin-peptide-stimulated leukocytes (Varga et al. 1990; Fülöp et al. 1992).

These results suggested an uncoupling of the elastin receptor from its normal intracellular transmission mechanism (Robert 1998).

3.3
Experiments on Lymphocytes

When human lymphocytes were isolated from tonsillar lymphoid tissue and cultured in presence of mitogens, the elastin receptor was progressively expressed on about 60% of cells, as shown by immunofluorescence and flow cytometry (Peterzsegi et al. 1997a,b). When elastin peptides were added to the culture, the expression of the receptor could be induced (increased) on helper (CD4) and memory (CD45 RO+) lymphocytes. In presence of low concentrations of K-elastin, cell proliferation was stimulated but also the secretion of a potent serine elastase, indistinguishable from PMN elastase (Peterszegi et al. 1997a; Péterszegi and Robert 1998; (Fig. 3A).

After further increase of elastin peptide concentration (above about $0.1\,\mathrm{mg\,ml^{-1}}$), increasing numbers of cells were found to take up the vital dye, Evans blue, indicating increasing cell death (Fig. 3B). Using the tunnel effect and electron microscopy, we could show that in presence of lower concentrations of K-elastin (0.1 to $1\,\mathrm{mg\,ml^{-1}}$) necrotic type of cell death predominated. At higher concentration of elastin peptides ($1-5\,\mathrm{mg\,ml^{-1}}$) apoptotic cell death could be observed (Peterszegi et al. 1999).

These results could be interpreted as the progressively increasing harmful effect of the overstimulated elastin receptor, presumably through the increasing release of serine proteases and free radicals.

Similar age-dependent modifications were observed on endothelial cells also. The vasodilatory effect of elastin peptides, shown to be NO˙-dependent (Faury et al. 1995), was maximal in young adult rats and decreased progressively with age (Faury et al. 1997).

We attributed this effect also to the progressive age-dependent uncoupling of this receptor from its transmission pathway. Free radicals and enzyme-mediated cell damage could be involved here also. In the case of simultaneous liberation of superoxide and nitric oxide, the highly toxic peroxynitrite anion ($ONOO^-$) could be formed. Its detoxification necessitates reduced glutathione, shown to decrease with age. PMN elastase was also shown to induce apoptosis in endothelial cells.

This age-dependent uncoupling of the elastin receptor is not specific for this receptor only. A number of receptors were shown to decrease or disappear with aging (for a review, see Robert 1998). Progressive loss of cell-matrix communication might well be an important factor in age-dependent loss of tissue structure and function. The role of similar or identical mechanisms was also shown to play a role in the progression of the atherosclerotic lesions (Peterszegi et al. 1997a; Robert et al. 1998).

Effect of elastin peptides (75 kD) on human lymphocyte cell proliferation and elastase production

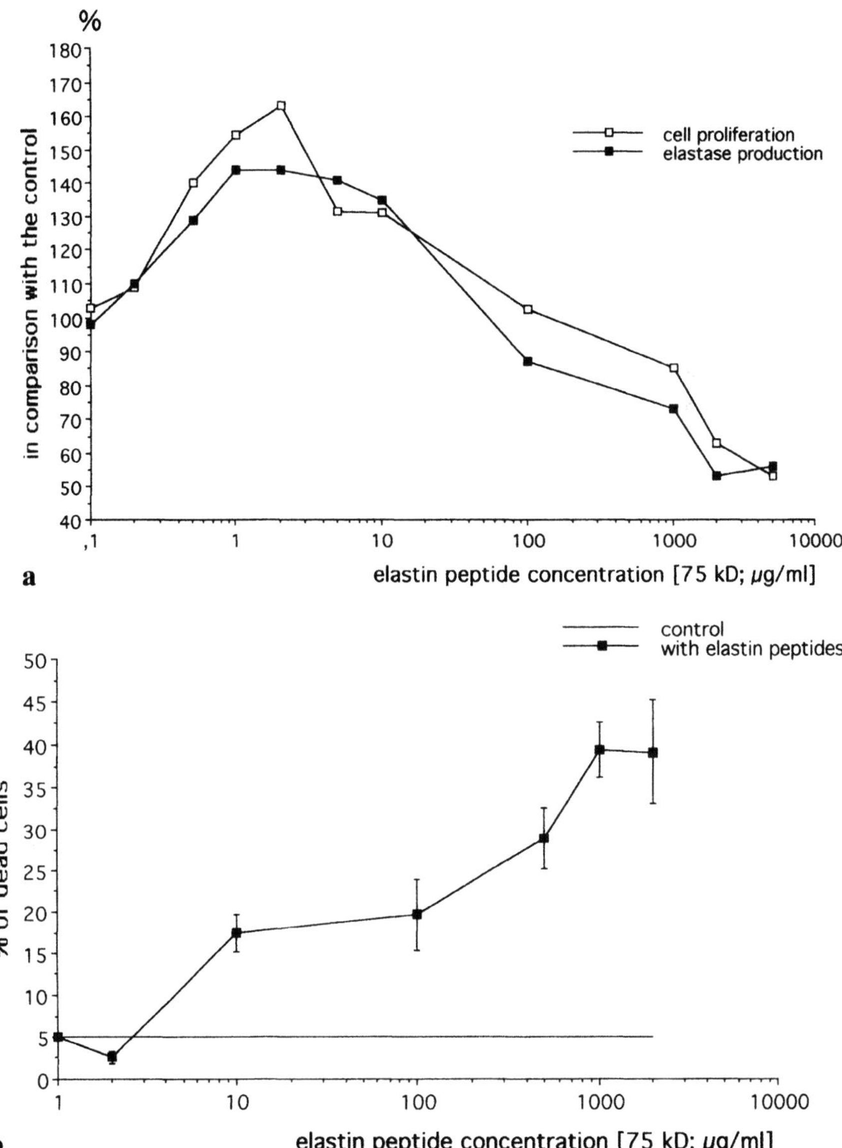

Fig. 3. a Effect of elastin peptides (log concentration on abscissa) on the proliferation (-□-) and elastase production (-■-) of activated human T-lymphocytes. The decrease in these curves above a critical elastin peptide concentration (~1–5 μgml^{-1}) is due to an increasing cell death, as shown in **b** using the uptake of a vital dye

4
Discussion and Conclusions

The two examples developed in this chapter clearly demonstrate the fact that the postsynthetic matrix is highly coupled to cells by receptors. These receptors transmit signals in both ways between cells and surrounding matrix components. Coupling can be direct by connecting ECM components to cell receptors or indirect when matrix glycoproteins mediate interactions. Fibronectin, in particular, was shown to mediate interactions between cells and other matrix components, collagen and proteoglycans in particular. Methods are now available to analyze in details the microarchitecture of the macromolecules in close connection with cell membranes. Rapid progress was made in the description of the intracellular transmission pathways between receptors and the cytoskeleton and the nucleus. Although we are far from understanding all the details of these transmission mechanisms, it is becoming clear that they differ with cell phenotype and also according to the surrounding macromolecules. Progress has to be made also on the detailed description of switch mechanisms which enable cells to receive signals from and send signals out to the matrix. The example of the elastin-laminin receptor demonstrates the differences between cell phenotype and signal transmission pathways. The intermediary steps were shown to be different between mononuclear cells and endothelial cells (Faury et al. 1998a; Fülöp et al. 1998). Another important fact demonstrated with this receptor was its progressive uncoupling during aging (Faury et al. 1997; Robert 1998). The reaction of this receptor with elastin peptides produced by the degradation of elastin fibers illustrates the potential harmful effect of the overcharging of this receptor by potential agonists. The progressive loss of normal receptor function and the potential harmful effects of uncoupled receptors may well represent the essential epigenetic mechanisms of aging.

References

Archilla-Marcos M, Robert L (1993) Control of the biosynthesis and excretion of the elastase-type protease of human skin fibroblasts by the elastin receptor. Clin Physiol Biochem 10:86–91

Balazs EA (ed) (1970) Chemistry and molecular biology of the intercellular matrix I-II-III: Academic Press, London

Bizbiz L, Alpérovitch A, Robert L, the EVA group (1997) Aging of the vascular wall: serum concentration of elastin peptides and elastase inhibitors in relation with cardiovascular risk factors. The EVA study. Atherosclerosis 131:73–78

Boyer B, Kern P, Fourtanier A, Labat-Robert J (1991) Age-dependent variations of the biosyntheses of fibronectin and fibrous collagens in mouse skin. Exp Gerontol 26:375–382

Defilippi P, Gismondi A, Santoni A, Tarone G (1997) Signal transduction by integrins. Springer, Berlin Heidelberg New York

De Melker AA, Sonnenberg A (1999) Integrins: alternative splicing as a mechanism to regulate ligand binding and integrin signaling events. Bioessays 21:499–509

Faury G, Ristori MT, Verdetti J, Jacob MP, Robert L (1995) Effect of elastin peptides on vascular tone. J Vasc Res 32:112–119

Faury G, Chabaud A, Ristori MT, Robert L, Verdetti J (1997) Effect of age on the vasodilatory action of elastin peptides. Mech Ageing Dev 95:31–42

Faury G, Usson Y, Robert-Nicoud M, Robert L, Verdetti J (1998a) Nuclear and cytoplasmic free calcium level changes induced by elastin peptides in human endothelial cells. Proc Natl Acad Sci 95:2967–2972

Faury G, Garnier S, Weiss AS, Wallach J, Fülöp T, Jacob MP, Mecham RP, Robert L, Verdetti J (1998b) Action of tropoelastin and synthetic elastin sequences on vascular tone and on free calcium level in human vascular endothelial cells. Circ Res 82:328–336

Fitton-Jackson S, Harkness RD, Partridge SM, Tristram GM (eds) (1965) Structure and function of connective and skeletal tissue. Auspices of NATO St Andrews, 15–25 June 1964. Butterworths, London

Flickinger KS, Carter WG, Culp LA (1992) Deficiency in integrin-mediated transmembrane signaling and microfilament stress fiber formation by aging dermal fibroblasts from normal and Down's syndrome patients. Exp Cell Res 203:466–475

Fornaro M, Languino LR (1997) Alternatively spliced variants: a new view of the integrin cytoplasmic domain. Matrix Biol 16:185–193

Frisch S, Francis H (1994) Disruption of epithelial cell-matrix interactions induces apoptosis. J Cell Biol 124:619–626

Frisch SM, Ruoslahti E (1997) Integrins and anoikis. Current opinion. Cell Biol 9:701–706

Fülöp T Jr, Jacob MP, Varga Zs, Foris G, Leovey A, Robert L (1986) Effect of elastin peptides on human monocytes: Ca^{2+} mobilization, stimulation of respiratory burst and enzyme secretion. Biochem Biophys Res Commun 141:92–98

Fülöp T, Wei SM, Robert L, Jacob MP (1990) Determination of elastin peptides in normal and arteriosclerotic human sera by ELISA. Clin Physiol Biochem 8:273–282

Fülöp T Jr, Barabas G, Varga Zs, Csongor J, Hauck M, Szücs S, Seres I, Mohacsi A, Kekessy D, Despont JP, Robert L, Penyige A (1992) Transmembrane signaling changes with aging. Ann New York Acad Sci 673:165–171

Fülöp T Jr, Jacob MP, Khalil A, Wallach J, Robert L (1998) Biological effects of elastin peptides. Pathol Biol 46:497–506

Ghuysen-Itard AF, Robert L, Gourlet V, Berr C, Jacob MP (1993) Loss of calcium-homeostatic mechanisms in polymorphonuclear leukocytes of demented and nondemented elderly subjects. Gerontology 39:163–169

Goldstein S (1990) Replicative senescence: the human fibroblast comes of age. Science 249:1129–1132

Groult V, Hornebeck W, Robert L, Pouchelet M, Jacob MP (1998) Interactions of elastin fibers with fibroblasts – a time-lapse cinemicrographic study. Pathol Biol 46:507–516

Hay ED (1991) Cell biology of extracellular matrix. Plenum Press, New York

Hineck A, Mecham RP (1990) Characterization and functional properties of the elastin receptor. In: Tamburro A (ed) Elastin: chemical and biological aspects. Galatina Congedo Editore Potenza, pp 369–381

Hornebeck W, Tixier JM, Robert L (1986) Inducible adhesion of mesenchymal cells to elastic fibers: elastonectin. Proc Natl Acad Sci 83:5517–5520

Hu Q, Moerman EJ, Goldstein S (1996) Altered expression and regulation of the $\alpha5\beta1$ integrin-fibronectin receptor lead to reduced amounts of functional heterodimer on the plasma membrane of senescent human diploid fibroblasts. Exp Cell Res 224:251–263

Hynes R0 (1987) Integrins: a family of cell surface receptors. Cell 48:549–554

Hynes RO (1990) Fibronectins. Springer Berlin Heidelberg, New York

Jacob MP, Hornebeck W (1985) Isolation and characterization of insoluble and kappa-elastins, In: Robert L, Moczar M, Moczar E (eds) Frontiers of matrix biology,vol 10. Karger, Basel, pp 92–123

Jacob MP, Fülöp T, Foris G, Robert L (1987) Effect of elastin peptides on ion fluxes in mononuclear cells, fibroblasts, and smooth muscle cells. Proc Natl Acad Sci 84:995–999

Kühn K (1997) Extracellular matrix constituents as integrin ligands. In: Eble JA, Kühn K (eds) Integrin-ligand interactions Springer Berlin, Heidelberg New York, pp 41–83

Labat-Robert J, Birembaut P, Robert L, Adnet JJ (1981a) Modification of fibronectin distribution pattern in solid human tumors. Diagn Histopathol 4:299–236

Labat-Robert J, Potazman JP, Derouette JC, Robert L (1981b) Age-dependent increase of human plasma fibronectin. Cell Biol Int 5:969–973

Mecham RP (1991) Elastin synthesis and fiber assembly. In: Pulmonary emphysema, the rationale for therapeutic intervention. Ann NY Acad Sci 624:137–146

Mecham RP, Hinek A, Griffin GL, Senior RM, Liotta LA (1989) The elastin receptor shows structural and functional similarities to the 67-kDa tumor cell laminin receptor. J Biol Chem 264:16652–16657

Mosher DF (1989) Fibronectin. Academic Press, London

Ooyama T, Fukuda K, Oda H, Nakamura H, Hikita Y (1987) Substratum-bound elastin peptide inhibits aortic smooth muscle cell migration in vitro. Arteriosclerosis 7:593–598

Perdomo JJ, Gounon P, Schaeverbeke M, Schaeverbeke J, Groult V, Jacob MP, Robert L (1994) Interaction between cells and elastin fibers: an ultrastructural and immunocytochemical study. J Cell Physiol 158:451–458

Peterszegi G, Robert L (1998) Cell death induced in lymphocytes expressing the elastin-laminin receptor by excess agonists: necrosis and apoptosis. Biomed Pharmacother 52:369–377

Peterszegi G, Mandet C, Texier S, Robert L, Bruneval P (1997a) Lymphocytes in human atherosclerotic plaque exhibit the elastin-laminin receptor: potential role in atherogenesis. Atherosclerosis 135:103–107

Peterszegi G, Texier S, Robert L (1997b) Human helper and memory lymphocytes exhibit an inducible elastin-laminin receptor. Int Arch Allergy Immunol 114:218–223

Peterszegi G, Texier S, Robert L (1999) Cell death by overload of the elastin-laminin receptor on human activated lymphocytes: protection by lactose and melibiose. Eur J Clin Invest 29:166–172

Pfaff M (1997) Recognition sites of integrins within their ligands. In: Eble JA, Kühn K (eds) Integrin-ligand interactions. Springer Berlin, Heidelberg New York, pp 101–121

Pierschbacher MD, Ruoslahti E (1984) Cell attachment activity of fibronectin can be duplicated by small synthetic fragments of the molecule. Nature 309:30–33

Rasoamanantena P, Labat-Robert J, Goldstein S (1993) Variations de la biosynthèse et quantification de l'ARNm de la fibronectine humaine au cours du vieillissement en culture. C R Soc Biol 187:238–246

Rasoamanantena P, Thweatt R, Labat-Robert J, Goldstein S (1994) Altered regulation of fibronectin gene expression in Werner syndrome fibroblasts Exp Cell Res 213:121–127

Robert L (1998) Mechanisms of aging of the extracellular matrix. Role of the elastin-laminin receptor, Novartis Price Lecture. Gerontology 44:307–317

Robert L, Hornebeck W (1986) Interaction between elastic fibers and cells. In: Labat-Robert J, Timpl R, Robert L (eds) Frontiers of matrix biology, vol 11. Karger, Basel, pp 58–77

Robert L, Robert AM, Jacotot B (1998) Elastin-elastase-atherosclerosis revisited. Atherosclerosis 140:281–295

Senior RM, Griffin GL, Mecham RP, Wrenn DS, Prasad KU, Urry DW (1984) Val-Gly-Val-Ala-Pro-Gly, a repeating peptide in elastin, is chemotactic for fibroblasts and monocytes. J Cell Biol 99:870–874

Varga Z, Kovacs EM, Paragh G, Jacob MP, Robert L, Fülöp T (1988) Effect of elastin peptides and N-formyl-methionyl-leucyl phenylalanine on cytosolic free calcium in polymorphonuclear leukocytes of healthy middle-aged and elderly subjects. Clin Biochem 21:127–130

Varga Z, Jacob MP, Robert L, Fulop T Jr (1989) Identification and signal transduction mechanism of elastin peptide receptor in human leukocytes. FEBS Lett 258:5–8

Varga Z, Jacob MP, Csongor J, Robert L, Leovey A, Fulop T Jr (1990) Altered phosphatidylinositol breakdown after K-elastin stimulation in MBIs of elderly. Mech Ageing Dev 52:61–70

Regulation of Gene Expression
by Changes in Cell Adhesion

Takejiro Kuzumaki[1]

1
Introduction

It is generally accepted that the interaction of cells with their extracellular environment is crucial for their coordinated functions and homeostasis. Differentiation and proliferation are highly coordinated events and represent the most basic behavior of cells. Biological signals to control cellular differentiation and proliferation are initiated by extracellular molecules, which can be divided into two classes based on their solubility. The soluble class includes hormones, growth factors, and cytokines, which circulate in the whole body or diffuse locally and bind to the specific receptors of target organs. The soluble factors are major and potent regulators, and their concentration in serum is finely controlled by integrated biological systems. The insoluble class is represented by (1) cell-cell interaction with adjacent cells and (2) cell adhesion to the extracellular matrix (ECM) (Gumbiner 1996). The insoluble factors are static regulators in comparison with the soluble factors, but have been found to play a critical role in the control of proliferation, differentiation, and morphogenesis.

Cadherins represent the major adhesion molecules in the adherence junction and mediate the interaction between adjacent cells. Cadherins form complexes with cytoplasmic plaque proteins, called catenins, as well as with the actin cytoskeleton. Other molecules involved in cell-cell interaction include desmogleins in the desmosomal junction and occludins in the occluding junction. These adhesion molecules play important roles in the maintenance of tissue structure, cell recognition, and cell-cell communications.

This chapter focuses mainly on the biological effects of cell adhesion to the ECM, especially on the regulation of proliferation and differentiation of cells. The ECM makes an intricate network of molecules, which is an important component of connective tissues and basement membrane. The ECM is composed of a variety of proteins and polysaccharides, for example, collagen, fibronectin, vitronectin, laminin, and proteoglycans, which are secreted by cells and assemble into fibrils or other complex macromolecular arrays. Cells adhere to this macromolecular matrix by anchoring to the components of ECM. These

[1] Department of Biochemistry, Yamagata University School of Medicine, Yamagata 990-9585, Japan

Progress in Molecular and Subcellular Biology, Vol. 25
A. Macieira-Coelho (Ed.)
© Springer-Verlag Berlin Heidelberg 2000

molecules bind to cell surface receptors, called integrins. Integrins are composed of α and β subunits selected from among 16 α and 8 β subunits, which heterodimerize to produce more than 20 different receptors. The integrin recognizes and binds to the specific ECM molecules; for example, α1β1 integrin recognizes collagen and laminin (Rosales and Juliano 1995). Integrins are localized in adhesion plaque (focal contact, focal adhesion), which is well developed in cultured fibroblasts. Actin-binding proteins, α-actinin, talin, vinculin, paxillin, and tensin, are capable of colocalizing with integrins in adhesion plaques. Cytoskeletal actin filaments link to integrins through these actin-binding proteins.

Adhesion plaques represent an important machinery regarding cell adhesion to the ECM, cell migration and motility, and the determination of cell shape. In addition to these functions, the biochemical signals are transduced from adhesion plaques when integrins bind to the ECM (Juliano and Haskill 1993; Clark and Brugge 1995; Gumbiner 1996). The signals are generated from integrin clustering by interaction of integrins with the ECM molecules. The activation of several protein tyrosine kinases which are localized in adhesion plaques is an essential event for integrin-dependent signaling; for example, the activation of focal adhesion kinase (FAK), Src, Fgr, Csk, and Syk. Among these, FAK appears to play a central role in integrin-mediated signal transduction. Autophosphorylated tyrosine residues of FAK proteins interact with the SH2-SH3-containing adaptor proteins, Grb2, followed by an interaction with the guanine nucleotide exchange factor, SOS, which activates ras-MAP kinase pathway. MAP kinase catalyzes the phosphorylation and subsequent activation of transcription factors, which regulate the transcription of several genes in the nucleus. In addition to the ras-MAP kinase pathway, several integrin-dependent signaling pathways have been proposed, which involves protein kinase C, phosphatidylinositol-3 kinase, phospholipase C, and the low molecular weight GTPases, Rho. Consequently, these signals generated by the binding of integrin to the ECM reach the nucleus and modulate transcription thus regulating the expression of several genes to control cell growth and differentiation.

In this chapter, the effect of cell adhesion to the ECM on the fibroblast cell cycle and cyclin A gene expression is described in the first section entitled the regulation of cell proliferation by cell adhesion. In the second section entitled the regulation of cell differentiation by cell adhesion, the regulation of albumin and β-casein gene expression in hepatocytes and mammary epithelial cells cultured with the ECM gel is described.

2
Regulation of Cell Proliferation by Cell Adhesion

In an *in vitro* culture system, most normal mammalian cells cannot proliferate without adhering to culture dishes. This phenomenon is referred to as anchorage-dependent growth (Stoker et al. 1968). When normal cells are transformed, they can grow without adhesion (anchorage-independent growth), which is taken as verification of the transformation of cells and is well correlated with tumorigenicity (Shin et al. 1975).

Three methods described below are commonly used to study the effect of cell adhesion to the ECM in terms of cellular proliferation. An agar suspension culture or agar plate culture is widely used to evaluate cell transformation (Macpherson and Montagnier 1964; Kuroki 1975). Transformed cells can grow and form visible colonies when cultured without adhesion on agar-coated dishes. A viscous methylcellulose-containing medium is also used to culture cells in suspension. Methylcellulose is not cytotoxic, and cells cultured in a methylcellulose-containing medium can be harvested for analysis of biochemical and biophysical events in the cells (Stoker et al. 1968). To alter the adhesivity of cells to plastic culture dishes gradually, poly (2-hydroxyethyl methacrylate) (polyHEMA) is a useful tool. By changing the concentration of polyHEMA used to coat the plastic culture dishes, the adhesiveness and morphology of the cells can be changed. Cell shape and adhesiveness were found to be tightly coupled to DNA synthesis and growth in nontransformed cells (Folkman and Moscona 1978).

Fibroblast cell lines are preferentially used to study the effect of cell adhesion on cell proliferation because data on the cell cycle events regulated by growth factors have accumulated and adhesion plaques are well developed in cultured fibroblasts. In the reports mentioned in this section, fibroblasts are the cells of choice for most experiments.

2.1
Regulation of the G_1 Phase

For the proliferation of cultured fibroblasts, two critical factors, growth factors in serum and cell adhesion to substratum (culture dishes), are required. Growth factors and cell adhesion are necessary for the progression of G_1 phase (Otsuka and Moskowitz 1975; Okuda and Kimura 1984; Han et al. 1993; Kuzumaki et al. 1996). Fibroblasts which are cultured in serum-deprived medium (without growth factors) become quiescent, which arrests the cell cycle at the G_0 state (quiescent state). When quiescent cells are stimulated with growth factors, they enter the cell cycle through the G_1 phase. Once cells enter the S phase beyond the restriction point located at the G_1/S boundary, the subsequent S, G_2, and M phases progress independently of growth factors and cell adhesion. Growth factors induce immediate-early response genes, c-fos, c-jun, junB, and c-myc, within several hours after stimulation of quiescent cells.

Cell adhesion is critical for the transition of G_1 to S phase (Guadagno and Assoian 1991; Kuzumaki et al. 1996). Kinetic studies indicate that cells are arrested at the late G_1 phase in suspension cultures and cell adhesion is required in the late G_1 phase, especially at the period of G_1/S transition. Even if cells are cultured without adhesion, c-fos and c-myc genes are induced by serum stimulation and the cell size increases in the G_1 phase. These results suggest that cell adhesion plays an important role in the restriction point control at the G_1/S boundary, not in the G_0/G_1 transition.

However, several studies have shown that the effect of cell adhesion is not limited to the events which occur in the G_1/S transition. When methylcellulose suspension-arrested cells attach to the culture dishes, the adhesion of these suspension-arrested cells rapidly induces the expression of immediate-response genes, c-fos, c-myc, actin, and pro-$\alpha 1$(I) collagen, in mouse fibroblasts (Dike and Farmer 1988; Dhawan et al. 1991). Increase in cytosolic calcium levels and inositol lipid turnover rate induced by platelet-derived growth factor is abolished in the case of suspension cultures of fibroblasts (Tucker et al. 1990; McNamee et al. 1993). Ornithine decarboxylase, which is preferentially expressed in the mid-G_1 phase, is also suppressed in the cells cultured on polyHEMA-coated dishes in nontransformed epithelial cells (Morrison and Seidel 1995). In CUA-1 fibroblasts, the expression of c-myc and c-ras genes has been reported to be suppressed in the cells cultured on polyHEMA-coated dishes (Farrell and Greene 1992). These reports indicate that cell adhesion is also necessary for the G_0/G_1 transition and the progression of mid-G_1 phase.

Cell spreading and cell attachment to the ECM with integrins act at different points in the cell cycle. When the primary hepatocytes were cultured on RGD peptide-coating dishes, integrins bind to the RGD peptides, but the cells do not spread. In these cells, junB is normally expressed but the cells can not enter the S phase (Hansen et al. 1994). When capillary endothelial cells adhere to low concentration of fibronectin or to small adhesive islands, cells do not spread and fail to progress through G_1 and enter S phase. This shape-dependent block in cell cycle progression correlates with a failure to increase cyclin D1 and to down-regulate the cell cycle inhibitor p27 (Huang et al. 1998). Cell spreading may be a critical factor for the transition of G_1 to S phase in adherent cells.

2.2
Molecular Basis for Regulation of the G_1/S Transition

The most critical events in progression of the cell cycle are the sequential activation of several types of cyclin-dependent kinases (CDK). For progression of the G_1 and S phases, activation of cdk4 and cdk2 is necessary (Sherr 1993). Cdk4 is activated by binding to cyclin D1 in the mid- and late-G_1 phase. Cdk2 is activated by binding to cyclin E at the G_1/S boundary, and is also activated by binding to cyclin A during the S phase. Their target molecule is an Rb protein encoded by the retinoblastoma-suppressor gene. The phosphorylation of Rb, which is cooperated by both cdk4 and cdk2, is one of the important events for entering the S phase from G_1. Cyclin-dependent kinase inhibitors (CDKI) have been identified as negative regulators to these CDK activities. One class of CDKI includes p21 and p27, which can bind to and inhibit both cdk4 and cdk2. Another class of CDKI includes p16, p15, p18, and p19, which bind to and inhibit cdk4 (Sherr 1994).

The effect of cell adhesion on CDKs and their related molecules in fibroblasts has been reported (summarized in Table 1). Guadagno et al. (1993) formed that the amounts of cdk2, cdc2, cyclin D1, and cyclin E proteins are not changed, but that the protein and mRNA of cyclin A are decreased in nonadherent mouse NIH 3T3 and rat NRK fibroblasts. Furthermore, fibroblasts can grow in suspension when the cyclin A-containing plasmid is transfected. The results indicate that cyclin A gene expression is critical for the control of growth by cell adhesion. In mouse NIH 3T3 fibroblasts, Schulze et al. (1996) showed that suppression of cyclin A expression in nonadherent cells is caused by an increase in p27 protein. The increase in p27 protein is not caused by the transcriptional activation of its gene but by the suppression of its degradation rate. In their study, p21 does not increase in nonadherent cells. They conclude that in suspension cells the increase in p27 protein inhibits the activities of cdk2/cyclin E and cdk4/cyclin D1, followed by the decrease in Rb-phosphorylation and cyclin A gene expression. Other reports show that levels of p21 as well as p27 also increase in nonadherent mouse BALB/c 3T3 and human fibroblasts, and the activity of cdk2/cyclin E is inhibited (Fang et al. 1996; Kuzumaki and Ishikawa 1997).

Cyclin D1 seems to be important for anchorage-dependent cell cycle regulation since in mouse NIH 3T3, rat1, and human skin fibroblasts, cyclin D1 is suppressed in nonadherent cells (Zhu et al. 1996; Resnitzky 1997; Schulze et al. 1998). In these studies, the increase in p27 and/or p21 levels and the suppression of cdk2/cyclin E activity were also observed in addition to the inhibition of cdk4/cyclin D1 activity. Resnitzky (1997) showed that the ectopic expression of cyclin D1, but not cyclin E, induces the anchorage-independent cell cycle progression. In addition, Schulze et al. (1996) also showed that the overexpression of cyclin D1 relieves the adhesion-dependent block of cyclin A transcription and allows DNA replication in suspension culture. However, other works showed that cyclin D1 expression is not affected by a change in adhesion status (Table 1).

Focal adhesion kinase (FAK) plays a role in p21 and cyclin D1 expression. The overexpression of wild-type FAK increases cyclin D1 and decreases p21 expression, and a dominant-negative mutant of FAK blocks the upregulation of cyclin D1 and increases p21 in fibroblasts (Zhao et al. 1998).

In other types of cells, the effect of cell adhesion on the regulation of cell cycle has also been reported. In murine myeloid progenitor 32D cells, the ectopic expression of chronic myelogeneous leukemia-specific fusion protein p210bcr/abl stimulates the expression of $\alpha5\beta1$ integrin and increases cell adhesion to fibronectin. In the transfectant, the G_i/S transit progresses more readily and cdk2/cyclin A activity is increased. The increase in cdk2/cyclin A activity does not correlate with cyclin A expression but, rather, with the decrease in p27 (Krämer et al. 1999). In human arterial smooth muscle cells, the proliferation rate is increased in the cells cultured on monomer collagen compared to the cells cultured on polymerized collagen. In the case of cells cultured on polymerized collagen, cyclin A expression and cdk2/cyclin E activity are

Table 1. Change of cyclin-dependent kinases and their related molecules in suspension culture of fibroblasts

Fibroblast cell line	Cyclin D1	Cyclin E	Cyclin A	cdk4/D1	cdk2/E	p21	p27	Rb-P	Reference
Mouse NIH3T3	N	N	D	–	–	–	–	–	Guadagano et al. (1993)
	N	N	D	D	D	N	I	D	Schulze et al. (1996)
	D	N	D	D	D	I	I	D	Zhu et al. (1996)
	–	N	D	–	D	–	–	D	Kang and Krauss (1996)
	D	N	D	D	D	–	I	D	Schulze et al. (1998)
Mouse BALB/c 3T3	N	N	D	–	D	I	I	–	Kuzumaki and Ishikawa (1997)
Rat NRK	N	N	D	–	–	–	–	–	Guadagano et al. (1993)
	N	N	D	N	N	N	N	N	Zhu et al. (1996)
	N	N	D	–	D	N	N	–	Carstens et al. (1996)
Rat PKC3-F4, ER1-1	N	N	D	–	D	–	N	D	Kang et al. (1996)
Rat1	D	N	D	–	D	N	I	–	Resnitzky et al. (1997)
Rat1a	N	N	D	–	–	–	–	–	Barrett et al. (1995)
Human skin	D	N	D	–	D	I	I	D	Zhu et al. (1996)
Human KD	N	N	–	N	D	I	I	–	Fang et al. (1996)

Abbreviations: N, no change; D, decreased; I, increased; –, not determined; cdk4/D1, kinase activity of cdk4/cyclinD1; cdk2/E, kinase activity of cdk2/cyclinE; Rb-P, phosphorylation of Rb protein.

suppressed, and both p21 and p27 are increased (only p27 binds to the cdk2 complex) (Koyama et al. 1996).

In summary: (1) In most reports, loss of cell adhesion upregulates p27 and/or p21, followed by the inhibition of cdk2/cyclin E activity and the suppression of Rb-phosphorylation. (2) In some reports, loss of cell adhesion downregulates cyclin D1, followed by the inhibition of cdk4/cyclin D1 activity. (3) In all studies performed with fibroblasts, the expression of cyclin A gene is suppressed by loss of cell adhesion.

2.3
Regulation of Cyclin A Gene Expression

As mentioned above, the biochemical targets of cell adhesion are CDKI (p21 and p27), cyclin D1, and cyclin A whose regulation of gene expression by cell adhesion has been well studied. The suppression of cyclin A gene expression by loss of cell adhesion is universally observed in most cell lines which have been examined and may be a final target of the anchorage-dependent cell cycle regulation. Luciferase assay using deletion mutants of the 5′-upstream region of human cyclin A gene indicates that the region lying between $-79 \sim +11\,$bp is essential for cell cycle regulation of cyclin A gene expression by growth factors (Henglein et al. 1994; Schulze et al. 1995). In this promoter region, several ciselements that play an important role in the cell cycle regulation of cyclin A gene expression have been identified (Fig.1). It has been reported that the GC-rich region located at around $-34\,$bp is necessary for the cell cycle regulation of cyclin A gene expression (Zwicker et al. 1995; Huet et al. 1996). The core sequence of this element is CGCGG, which is designated CDE (cell cycle dependent element) or CCRE (cell cycle responsive element). This element is

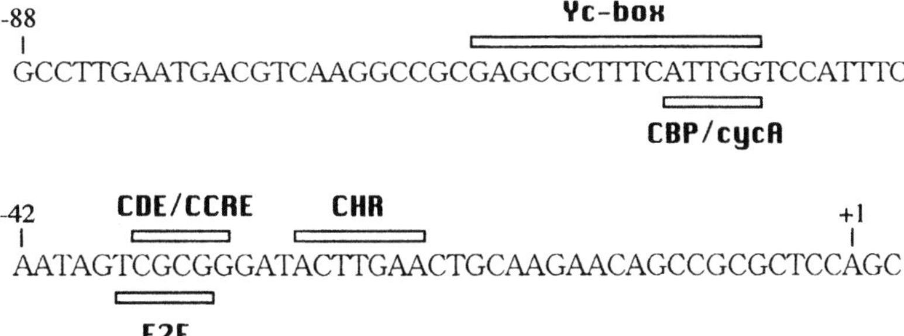

Fig. 1. The promoter region in cyclin A gene implicated in the anchorage-dependent transcription. The promoter sequence ($-88 \sim +3\,$bp) in human cyclin A gene and the ciselements implicated in the regulation of cell cycle- and anchorage-dependent transcription of cyclin A gene are shown. CDE, cell cycle dependent element; CCRE, cell cycle responsive element; CHR, cell cycle genes homology region; CBP/cycA, CCAAT binding protein for cyclin A gene. (After Zwicker et al. 1995)

conserved among cdc25C and cdc2 genes, which are periodically expressed in the cell cycle. This is a suppressive element and is required for arresting the cell cycle at the G_0 state (quiescent state by serum deprivation). The transcription factors that recognize this element have not yet been identified. CHR (cell cycle genes homology region), just downstream of the CDE/CCRE element, has also been implicated in the cell cycle regulation of cyclin A gene. In anchorage-dependent transcription of the cyclin A gene, the CDE/CCRE element is required for the suppression of cyclin A gene in suspension cells (Philips et al. 1999). If the CDE/CCRE element is mutated, suppression of cyclin A transcription is relieved in nonadherent cells. In addition, the CDE/CCRE-dependent suppression of cell adhesion requires a functional Rb protein.

The transcription factor E2F is a key factor for the G_1/S transition. A non-phosphorylated Rb can bind to and inactivate E2F in the G_1 phase. Both cdk4 and cdk2 catalyze the phosphorylation of Rb, which dissociates E2F from Rb. Free E2F activates the transcription of S phase-specific genes such as the dihydrofolate reductase and DNA polymerase-α genes. The E2F-binding site is also found in the cyclin A promoter region. The core sequence of E2F-binding element is TCGCG, which is located at around -35 bp in the cyclin A gene. The E2F-binding site is also required for the cell cycle regulation and the anchorage-dependent regulation of cyclin A gene (Schulze et al. 1995, 1996). If the E2F-binding site is mutated, the suppression of cyclin A expression is relieved in nonadherent cells and the anchorage-dependent transcription is also abolished. It should be noted that the E2F-binding site and CDE/CCRE element are located at the same position and share 80% of the core sequence (Fig. 1). These findings raise the possibility that the CDE/CCRE-binding factor and E2F are switched to bind to their elements for regulation of cyclin A transcription.

Another transcription factor implicated in the cell cycle regulation of cyclin A gene is the transcription factor NF-Y. NF-Y is one of the CCAAT-binding factors and has been described to mediate the serum-dependent transcriptional activation of human thymidine kinase gene at the G_1/S boundary (Chang and Liu 1994). The NF-Y binding site is located in the promoter region of cyclin A gene (Yc-box, Fig. 1). Krämer et al. (1996) showed that a novel CCAAT-binding protein designated CBP/cycA (CCAAT-binding protein for cyclin A gene) is required for the anchorage-dependent transcription of cyclin A gene. The CBP/cycA binds to the element present in the Yc-box (Fig. 1). Transcription of the cyclin A gene is not activated in nonadherent cells, and mutation of the CBP/cycA-binding site abolishes the activation of cyclin A gene transcription triggered by growth factors in adherent cells. This finding suggests that adhesion supports the binding of CBP/cycA to the cyclin A promoter. The components of CBP/cycA have been found to be NF-YA and NY-YB, which are the subunits of NF-Y (Krämer et al. 1997).

On the other hand, it has been reported that c-myc is involved in the anchorage-dependent regulation of cyclin A gene. However, the direct interaction of c-myc to the promoter region of cyclin A gene has not been demonstrated (Barrett et al. 1995).

In summary, the model of anchorage-dependent regulation of cyclin A gene expression is as follows: loss of cell adhesion leads to the binding of CDE/CCRE-binding factors to the CDE/CCRE element, which prevents the binding of E2F and CBP/cycA (NF-Y) to their specific binding sites (cyclin A transcription is inactivated). In adherent cells, the CDE/CCRE-binding factors are removed from the element, which allows E2F and CBP/cycA to bind to their elements, and this switching activates the transcription of cyclin A gene. A functional Rb protein may be involved in this switching of transcription factors. It is important to state that cell adhesion works coordinately with growth factors in activating the cyclin A gene since cyclin A is not expressed if growth factors are not present in adherent cells.

3
Regulation of Cell Differentiation by Cell Adhesion

The ECM is one of the important factors in regulation of cellular differentiation. When cells are cultured on plastic culture dishes or on thin layer collagen-coated dishes, spreading occurs and the differentiation phenotype is abolished in most mammalian cells. When cells are cultured by overlay with collagen gel or EHS gel (reconstituted gel extracted from Engelbreth-Holm-Swarm mouse tumor), the full surface of cells is in contact with the ECM, the cells become thick and appear cuboidal in shape, and the differentiated phenotype is maintained. The regulation of albumin gene expression in hepatocytes and the regulation of β-casein gene expression in mammary epithelial cells are described in this section as representative models of differentiation by the ECM gel.

In contrast to hepatocytes and mammary epithelial cells, it has been observed that differentiation of keratinocytes is inhibited by cell adhesion to the ECM; loss of cell-substratum interaction appears to be the trigger for terminal differentiation of keratinocytes (Green 1977), the differentiation being inhibited by fibronectin (not collagen I, collagen IV, and laminin) (Adams and Watt 1989). The epidermis is made up of multiple layers of keratinocytes with the basal layer in contact with the basement membrane which contains a range of ECM molecules including type IV collagen, laminin, fibronectin, and heparan sulfate proteoglycan. Terminal differentiation of keratinocytes is suppressed in this layer where the cells are highly proliferative. As the cells move away from the basement membrane and up to the suprabasal layers, they gradually acquire the differentiated phenotype of the epidermis. The basement membrane might inhibit keratinocytes from terminal differentiation, thus the inhibition of keratinocyte differentiation by adhesion to the ECM seems physiological.

A representative example of the induction of cellular functions triggered by adhesive interaction with the matrix are the monocytes. To extravasate into sites of tissue injury, infection, and tissue remodeling, monocytes must adhere to and migrate over the surface of vascular and connective tissue cells and the ECM which surrounds these cells. After adherence to plastic dishes in an experimental culture system, an impressive number of genes associated with the inflammatory response are expressed, for example, IL-1β, TNFα, and IL-8. In the 5′-upstream region of these inducible genes, the transcription factor NF-kB binding site is commonly found (Juliano and Haskill 1993). The mRNA stability element, AUUUA, is also commonly found in the 3′-untranslated region of these mRNAs. Adhesion may be sufficient for activation of NF-kB to induce these genes and to suppress the degradation rates of the mRNAs of these genes in monocytes.

3.1
Differentiation of Hepatocytes and Regulation of Albumin Gene Expression

The primary culture of freshly isolated adult rat hepatocytes is a widely used system for studies of liver-specific function. The hepatocytes quickly lose their differentiated characteristics after being plated on tissue culture plastic; albumin gene transcription, which is the most commonly used marker for hepatocyte differentiation, decreases to an undetectable level after only 12h in culture. By precoating plastic dishes with a thin layer of type I collagen, albumin expression also decreases. When the EHS gel is used as a substratum, the expression of albumin gene is maintained (Ben-Ze'ev et al. 1988). The EHS gel is composed of laminin, type IV collagen, proteoglycan, and other molecules, and laminin is most effective for albumin induction among its components (Caron 1990). The sandwich culture of type I collagen (TIC) gel (cells are sandwiched by two TIC gel layers) is more effective than a single layer of TIC gel in maintaining the albumin expression until 6–7 weeks after plating (Dunn et al. 1992). The EHS and TIC gel cultures are also effective in inducing other liver-specific genes such as α1-antitrypsin, α1-inhibitor III, transferrin, tyrosine aminotransferase, phosphoenolpyruvate carboxykinase, aldolase B, and cytochrome P450 (Ben-Ze'ev et al. 1988; Ouchi et al. 1998; Gómez-Lechón et al. 1998).

The transcriptional control of albumin gene has been studied. The enhancer region located at around −10 kbp along with its promoter region is necessary for the liver-specific regulation of albumin gene expression (Pinkert et al. 1987). In this albumin enhancer region, 11 ciselements have been identified and designated eA ~ eK (Liu et al. 1991). The enhancement of albumin expression in the hepatocytes cultured in the TIC gel is mediated by eG and eH elements (DiPersio et al. 1991). The sequences of eG and eH elements are shown in Fig. 2a. The eG element contains an HNF1/HNF3-binding site-like sequence and the eG element-binding factor appears identical to HNF3. The eH element contains the TGTTTGC sequence which is found at regulatory sites of the albumin promoter, of the hepatitis B virus enhancer, and of other hepatic genes

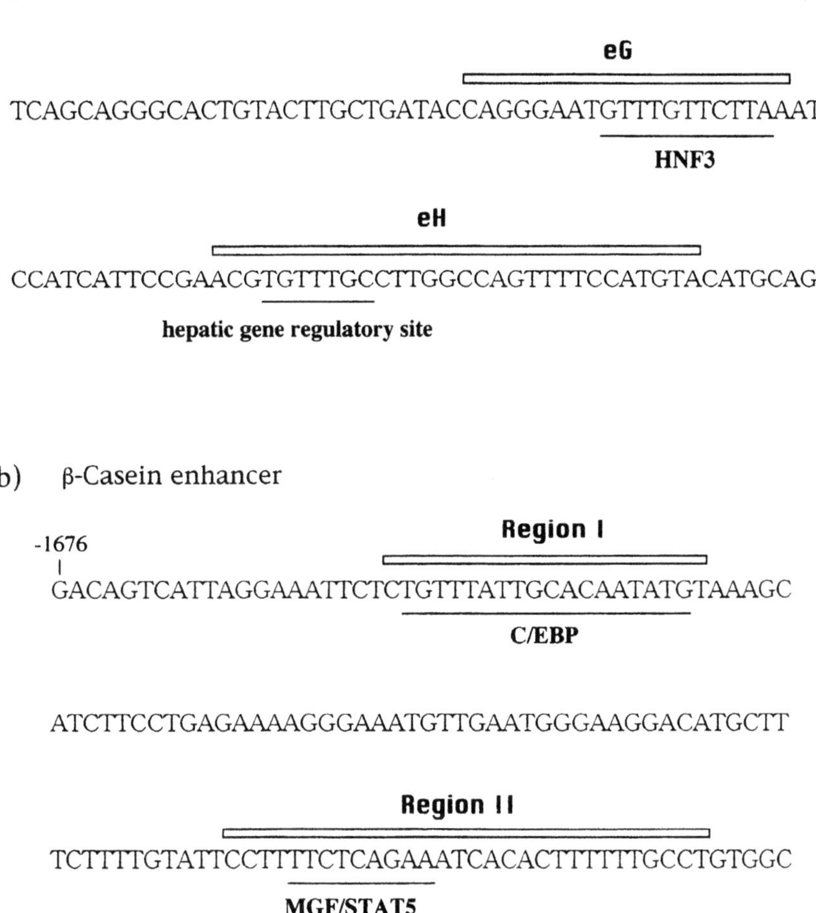

Fig. 2a,b. The enhancer regions in albumin and β-casein genes implicated in the extracellular matrix-dependent transcription. **a** The enhancer sequence in mouse albumin gene located at around −11 kbp and the ciselements implicated in the regulation of extracellular matrix-dependent transcription of albumin gene are shown. (After Liu et al. 1991). **b** The enhancer sequence in bovine β-casein gene (−1676 ~ −1517 bp), which is designated BCE-1 (bovine casein element-1), and the speculated ciselements implicated in the regulation of extracellular matrix-dependent transcription of β-casein gene are shown. MGF, mammary gland factor. (After Myers et al. 1998)

(it is tentatively referred to the hepatic gene regulatory site). The eH element-binding factor has not yet been identified.

To date, the HNF, C/EBP, and PAR families have been identified as liver-enriched transcription factors, and appear to be involved in liver-specific gene expression (Cereghini 1996). The expression of these transcription factors is regulated by the ECM. HNF3α, HNF1, and HNF4 are induced by the cultivation of cells with EHS gel (DiPersio et al. 1991; Nagaki et al. 1995; Oda et al. 1995). C/EBPα expression is also upregulated in the hepatocytes cultured with EHS gel (Rana et al. 1994). In the hepatocytes cultured with ECM gel, the differentiation phenotype is maintained and, in contrast, cell growth is suppressed. Consistent with this result, the expression of immediate-early response genes (c-jun, junB, c-myc, and c-fos) and cytoskeletal genes (actin, tubulin, and cytokeratin) are suppressed in the culture with EHS gel (Ben-Ze'ev et al. 1988; Rana et al. 1994; Gómez-Lechón et al. 1998). The ECM may have the master key for switching of cell growth and differentiation, via the regulation of expression of liver-enriched and growth-related transcription factors.

3.2
Differentiation of Mammary Epithelial Cells and Regulation of β-Casein Gene Expression

The mammary gland is one of the few adult tissues that undergoes dramatic and continuous morphological and functional changes after puberty. The differentiation process of mammary epithelial cells can be unambiguously followed by observing the profile of milk protein expression. In rodents, two classes of milk proteins can be defined, based on the time of their appearance in the gland. A representative of the first class of such proteins is β-casein. The other class of milk proteins is expressed later in pregnancy and whey acidic protein is an example (Lin and Bissell 1993).

Differentiation of mammary epithelial cells is induced by culturing with the EHS gel. β-Casein gene expression, which is a differentiation marker of mammary epithelial cells, is independent of cell-cell interaction with adjacent cells; the basement membrane, however, plays a central role in the expression of the gene mediated by laminin which is the major component of the basement membrane (Streuli et al. 1991). Type IV collagen and fibronectin are not able to induce β-casein expression (Streuli et al. 1995). For the ECM-dependent β-casein expression, the rounding and clustering of the cells (morphological change) are required, in addition to the biochemical signals associated with β1 integrin clustering (Roskelley et al. 1994). It should be noted that β-casein gene activation by the ECM requires the presence of prolactin.

CAT assay using deletion mutants of the 5'-upstream region of β-casein gene indicates that the region lying between −1517 and −1676 bp is necessary for the ECM responsiveness of bovine β-casein gene. This stretch of DNA acts as a transcriptional enhancer, and is designated BCE-1 (bovine casein element-1) (Schmidhauser et al. 1992). Mutation analysis of BCE-1 showed that two elements (region I and region II) are essential for the response to ECM (Myers et al. 1998). The consensus sequences of C/EBP and MGF/STAT5 are found within these regions (Fig. 2b). The electrophoretic mobility shift analysis indicates that C/EBP-β and STAT5 can bind to these elements; however, the binding activity to the elements of these transcription factors is not changed by the ECM. It is interestingly noticed that histone deacetylase inhibitors (sodium butylate and trichostatin A) are sufficient to induce transcription of integrated BCE-1 in the absence of ECM. Since histone acetylation is important for relaxing the chromatin structure, this suggests that the ECM induces a change in the β-casein gene which allows interaction between the transcription factors and the elements located in chromosomal DNA.

4
Conclusion

The ECM regulates the expression of several genes which play a critical role in the control of proliferation and differentiation of cells. This chapter provides an overview of the regulatory mechanism of gene expression by cell adhesion to the ECM in some experimental model systems using fibroblasts, hepatocytes, and mammary epithelial cells.

To date, the regulatory mechanisms of gene expression by the soluble factors have been extensively studied. Peptide hormones bind the specific receptors on the cell surface and generate biochemical signals that modulate the activity of transcription factors. Sometimes the signal transduction pathways are clear and simple. For example, glucagon binds to the receptors and increases intracellular cAMP. Increased cAMP activates protein kinase A, which phosphorylates CREB (CRE-binding protein). Activated CREB binds to the consensus CRE (cAMP-responsive element) sequence, thus activating glucagon-responsive genes. Another class of hormones, such as glucocorticoid, estrogen, and thyroxine, bind to the specific nuclear receptors. The activated receptors function as transcription factors and bind to the hormone-responsive elements; for example, glucocorticoid receptor binds to the consensus GRE (glucocorticoid-responsive element) sequence and activates glucocorticoid-responsive genes.

In comparison with the action of soluble factors, the molecular mechanism of regulation of gene expression by cell adhesion is more complicated and ambiguous. Cell adhesion and the soluble factors have been found to share the common transcription factors to regulate the cyclin A and albumin gene expression. Several reports have concluded that cell adhesion is able to biologically function only when the soluble factors are present. The transcription of cyclin A gene cannot be activated in adherent fibroblasts if growth factors are not present. Albumin and β-casein genes also are not expressed in the cultures with ECM gel, if hormones including prolactin are not present. Since cell adhesion works coordinately with the soluble factors, the effect of cell adhesion for gene expression cannot be distinguished from the action of soluble factors. Cell adhesion may function as a basic machinery, in order that the soluble factors may act completely.

Another difficult problem is that cell adhesion induces morphological changes in cells, in addition to generating the biochemical signals.

Morphological change has been found to be a critical factor in the function of cells in fibroblasts, hepatocytes, mammary epithelial cells, and capillary endothelial cells, for example, cell spreading is critical for the G_1/S phase transition of the fibroblast cell cycle, and the rounding of mammary epithelial cells is required for the ECM-dependent β-casein gene expression (DiPersio et al. 1991; Hansen et al. 1994; Roskelley et al. 1994; Morrison and Seidel 1995; Huang et al. 1998). However, it is not known how the cell shape affects the biochemical events in the cells.

For the reasons mentioned above, it may be difficult to identify the consensus sequence of an "extracellular matrix-responsive element"; however, it has recently proposed that cell adhesion affects chromatin structure via histone acetylation (Myers et al. 1998). It is possible that the chromatin restructuring is the universal mechanism of cell adhesion in regulating gene expression. Further study to understand the molecular mechanism of gene regulation by cell adhesion is required, which will, in turn, open new windows to understanding the mechanism of proliferation, differentiation, and apoptosis of cells.

References

Adams JC, Watt FM (1989) Fibronectin inhibits the terminal differentiation of human keratinocytes. Nature 340:307–309

Barrett JF, Lewis BC, Hoang AT, Alvarez RJ Jr, Dang CV (1995) Cyclin A links c-myc to adhesion-independent cell proliferation. J Biol Chem 270:15923–15925

Ben-Ze'ev A, Robinson GS, Bucher NLR, Farmer SR (1988) Cell-cell and cell-matrix interactions differentially regulate the expression of hepatic and cytoskeletal genes in primary cultures of rat hepatocytes. Proc Natl Acad Sci USA 85:2161–2165

Caron JM (1990) Induction of albumin gene transcription in hepatocytes by extracellular matrix proteins. Mol Cell Biol 10:1239–1243

Carstens C, Krämer A, Fahl WE (1996) Adhesion-dependent control of cyclin E/cdk2 activity and cell cycle progression in normal cells but not in Ha-ras transformed NRK cells. Exp Cell Res 229:86–92

Cereghini S (1996) Liver-enriched transcription factors and hepatocyte differentiation. FASEB J 10:267–282

Chang Z, Liu C (1994) Human thymidine kinase CCAAT-binding protein is NF-Y, whose A subunit expression is serum-dependent in human IMR-90 diploid fibroblasts. J Biol Chem 269:17893–17898

Clark EA, Brugge JS (1995) Integrins and signal transduction pathways: the road taken. Science 268:233–239

Dhawan J, Lichtler AC, Rowe DW, Farmer SR (1991) Cell adhesion regulates pro-α1(I) collagen mRNA stability and transcription in mouse fibroblasts. J Biol Chem 266:8470–8475

Dike LE, Farmer SR (1988) Cell adhesion induces expression of growth-associated genes in suspension-arrested fibroblasts. Proc Natl Acad Sci USA 85:6792–6796

DiPersio CM, Jackson DA, Zaret KS (1991) The extracellular matrix coordinately modulates liver transcription factors and hepatocyte morphology. Mol Cell Biol 11:4405–4414

Dunn JCY, Tompkins RG, Yarmush ML (1992) Hepatocytes in collagen sandwich: evidence for transcriptional and translational regulation. J Cell Biol 116:1043–1053

Fang F, Orend G, Watanabe N, Hunter T, Ruoslahti E (1996) Dependence of cyclin E-CDK2 kinase activity on cell anchorage. Science 271:499–502

Farrell RE Jr, Greene JJ (1992) Regulation of c-myc and c-Ha-ras oncogene expression by cell shape. J Cell Phys 153:429–435

Folkman J, Moscona A (1978) Role of cell shape in growth control. Nature 273:345–349

Gómez-Lechón MJ, Jover R, Donato T, Ponsoda X, Rodriguez C, Stenzel KG, Klocke R, Paul D, Guillén I, Bort R, Castell JV (1998) Long-term expression of differentiated functions in hepatocytes cultured in three-dimentional collagen matrix. J Cell Phys 177:553–562

Green H (1977) Terminal differentiation of cultured human epidermal cells. Cell 11:405–416

Guadagno TM, Assoian RK (1991) G1/S control of anchorage-independent growth in the fibroblast cell cycle. J Cell Biol 115:1419–1425

Guadagno TM, Ohtsubo M, Roberts JM, Assoian RK (1993) A link between cyclin A expression and adhesion-dependent cell cycle progression. Science 262:1572–1575

Gumbiner BM (1996) Cell adhesion: the molecular basis of tissue architecture and morphogenesis. Cell 84:345–357

Han EK, Guadagno TM, Dalton SL, Assoian RK (1993) A cell cycle and mutational analysis of anchorage-independent growth: cell adhesion and TGF-β1 control G1/S transit specifically. J Cell Biol 122:461–471

Hansen LK, Mooney DJ, Vacanti JP, Ingber DE (1994) Integrin binding and cell spreading on extracellular matrix act at different points in the cell cycle to promote hepatocyte growth. Mol Biol Cell 5:967–975

Henglein B, Chenivesse X, Wang J, Eick D, Bréchot C (1994) Structure and cell cycle-regulated transcription of the human cyclin A gene. Proc Natl Acad Sci USA 91:5490–5494

Huang S, Chen CS, Ingber DE (1998) Control of cyclin D1, p27Kip1, and cell cycle progression in human capillary endothelial cells by cell shape and cytoskeletal tension. Mol Biol Cell 9:3179–3193

Huet X, Rech J, Plet A, Vié A, Blanchard JM (1996) Cyclin A expression is under negative transcriptional control during the cell cycle. Mol Cell Biol 16:3789–3798

Juliano RL, Haskill S (1993) Signal transduction from the extracellular matrix. J Cell Biol 120:577–585

Kang J, Krauss RS (1996) Ras induces anchorage-independent growth by subverting multiple adhesion-regulated cell cycle events. Mol Cell Biol 16:3370–3380

Koyama H, Raines EW, Bornfeldt KE, Roberts JM, Ross R (1996) Fibrillar collagen inhibits arterial smooth muscle proliferation through regulation of cdk2 inhibitors. Cell 87:1069–1078

Krämer A, Carstens C, Fahl WE (1996) A novel CCAAT-binding protein necessary for adhesion-dependent cyclin A transcription at the G1/S boundary is sequestered by a retinoblastoma-like protein in G0. J Biol Chem 271:6579–6582

Krämer A, Carstens C, Wasserman WW, Fahl WE (1997) CBP/cycA, a CCAAT-binding protein necessary for adhesion-dependent cyclin A transcription, consists of NF-Y and a novel Mr 115000 subunit. Cancer Res 57:5117–5121

Krämer A, Hörner S, Willer A, Fruehauf S, Hochhaus A, Hallek M, Hehlmann R (1999) Adhesion to fibronectin stimulates proliferation of wild-type and bcr/abl-transfected murine hematopoietic cells. Proc Natl Acad Sci USA 96:2087–2092

Kuroki T (1975) Agar plate culture and Lederberg-style replica plating of mammalian cells. Methods Cell Biol 9:157–178

Kuzumaki T, Ishikawa K (1997) Loss of cell adhesion to substratum up-regulates p21Cip1/WAF1 expression in BALB/c 3T3 fibroblasts. Biochem Biophys Res Commun 238:169–172

Kuzumaki T, Matsuda A, Ito K, Ishikawa K (1996) Cell adhesion to substratum and activation of tyrosine kinases are essentially required for G1/S phase transition in BALB/c 3T3 fibroblasts. Biochim Biophys Acta 1310:185–192

Lin CQ, Bissell MJ (1993) Multi-faceted regulation of cell differentiation by extracellular matrix. FASEB J 7:737–743

Liu J, DiPersio CM, Zaret KS (1991) Extracellular signals that regulate liver transcription factors during hepatic differentiation in vitro. Mol Cell Biol 11:773–784

Macpherson I, Montagnier L (1964) Agar suspension culture for the selective assay of cells transformed by polyoma virus. Virology 23:291–294

McNamee HP, Ingber DE, Schwartz MA (1993) Adhesion to fibronectin stimulates inositol lipid synthesis and enhances PDGF-induced inositol lipid breakdown. J Cell Biol 121:673–678

Morrison RF, Seidel ER (1995) Cell spreading and the regulation of ornithine decarboxylase. J Cell Sci 108:3787–3794

Myers CA, Schmidhauser C, Mellentin-Michelotti J, Fragoso G, Roskelley CD, Casperson G, Mossi R, Pujuguet P, Hager G, Bissell MJ (1998) Characterization of BCE-1, a transcriptional enhancer regulated by prolactin and extracellular matrix and modulated by the state of histone acetylation. Mol Cell Biol 18:2184–2195

Nagaki M, Shidoji Y, Yamada Y, Sugiyama A, Tanaka M, Akaike T, Ohnishi H, Moriwaki H, Muto Y (1995) Regulation of hepatic genes and liver transcription factors in rat hepatocytes by extracellular matrix. Biochem Biophys Res Commun 210:38–43

Oda H, Nozawa K, Hitomi Y, Kakinuma A (1995) Laminin-rich extracellular matrix maintains high level of hepatocyte nuclear factor 4 in rat hepatocyte culture. Biochem Biophys Res Commun 212:800–805

Okuda A, Kimura G (1984) Control in previous and present generations of preparation for entry into S phase and the relationship to resting state in 3Y1 rat fibroblastic cells. Exp Cell Res 155:24–32

Otsuka H, Moskowitz M (1975) Arrest of 3T3 cells in G1 phase in suspension culture. J Cell Phys 87:213–220

Ouchi H, Otsu K, Kuzumaki T, Iuchi Y, Ishikawa K (1998) Synergistic induction by collagen and fibronectin of liver-specific genes in rat primary cultured hepatocytes. Arch Biochem Biophys 358:58–62

Philips A, Huet X, Plet A, Rech J, Vié A, Blanchard JM (1999) Anchorage-dependent expression of cyclin A in primary cells requires a negative DNA regulatory element and a functional Rb. Oncogene 18:1819–1825

Pinkert CA, Ornitz DM, Brinster RL, Palmiter RD (1987) An albumin enhancer located 10 kb upstream functions along with its promoter to direct efficient, liver-specific expression in transgenic mice. Gene Dev 1:268–276

Rana B, Mischoulon D, Xie Y, Bucher NLR, Farmer SR (1994) Cell-extracellular matrix interactions can regulate the switch between growth and differentiation in rat hepatocytes: reciprocal expression of C/EBPα and immediately-early growth response transcription factors. Mol Cell Biol 14:5858–5869

Resnitzky D (1997) Ectopic expression of cyclin D1 but not cyclin E induces anchorage-independent cell cycle progression. Mol Cell Biol 17:5640–5647

Rosales C, Juliano RL (1995) Signal transduction by cell adhesion receptors in leukocytes. J Leuko Biol 57:189–198

Roskelley CD, Desprez PY, Bissell MJ (1994) Extracellular matrix-dependent tissue-specific gene expression in mammary epithelial cells requires both physical and biochemical signal transduction. Proc Natl Acad Sci USA 91:12378–12382

Schmidhauser C, Casperson GF, Myers CA, Sanzo KT, Bolten S, Bissell MJ (1992) A novel transcriptional enhancer is involved in the prolactin- and extracellular matrix-dependent regulation of β-casein gene expression. Mol Biol Cell 3:699–709

Schulze A, Zerfass K, Spitkovsky D, Middendorp S, Berges J, Helin K, Jansen-Dürr P, Henglein B (1995) Cell cycle regulation of the cyclin A gene promoter is mediated by a variant E2F site. Proc Natl Acad Sci USA 92:11264–11268

Schulze A, Zerfass-Thome K, Berges J, Middendorp S, Jansen-Dürr P, Henglein B (1996) Anchorage-dependent transcription of the cyclin A gene. Mol Cell Biol 16:4632–4638

Schulze A, Mannhardt B, Zerfass-Thome K, Zwerschke W, Jansen-Dürr P (1998) Anchorage-independent transcription of the cyclin A gene induced by the E7 oncoprotein of human papillomavirus type 16. J Virol 72:2323–2334

Sherr CJ (1993) Mammalian G1 cyclins. Cell 73:1059–1065

Sherr CJ (1994) G1 phase progression: cycling on cue. Cell 79:551–555

Shin S, Freedman VH, Risser R, Pollack R (1975) Tumorigenicity of virus-transformed cells in nude mice is correlated specifically with anchorage independent growth in vitro. Proc Natl Acad Sci USA 72:4435–4439

Stoker M, O'Neill C, Berryman S, Waxman V (1968) Anchorage and growth regulation in normal and virus-transformed cells. Int J Cancer 3:683–693

Streuli CH, Bailey N, Bissell MJ (1991) Control of mammary epithelial differentiation: basement membrane induces tissue-specific gene expression in the absence of cell-cell interaction and morphological polarity. J Cell Biol 115:1383–1395

Streuli CH, Schmidhauser C, Bailey N, Yurchenco P, Skubitz APN, Roskelley C, Bissell MJ (1995) Laminin mediates tissue-specific gene expression in mammary epithelia. J Cell Biol 129:591–603

Tucker RW, Meade-Cobun K, Ferris D (1990) Cell shape and increased free cytosolic calcium $[Ca^{2+}]i$ induced by growth factors. Cell Calcium 11:201–209

Zhao J, Reiske H, Guan J (1998) Regulation of the cell cycle by focal adhesion kinase. J Cell Biol 143:1997–2008

Zhu X, Ohtsubo M, Böhmer RM, Roberts JM, Assoian RK (1996) Adhesion-dependent cell cycle progression linked to the expression of cyclin D1, activation of cyclin E-cdk2, and phosphorylation of the retinoblastoma protein. J Cell Biol 133:391–403

Zwicker J, Lucibello FC, Wolfraim LA, Gross C, Truss M, Engeland K, Müller R (1995) Cell cycle regulation of the cyclin A, cdc25C and cdc2 genes is based on a common mechanism of transcriptional repression. EMBO J 14:4514–4522

Expression of Liver Specific-Genes in Hepatocytes Cultured in Collagen Gel Matrix

Maria José Gómez-Lechón[1], Ramiro Jover[1], Teresa Donato[1], Xavier Ponsoda[2], and José V. Castell[1]

1
Introduction

It is wellknown that in conventional cultures of hepatocytes the expression of liver-specific genes declines dramatically after a short time in culture, and this precedes the gradual loss of drug metabolizing capability and cell functionality, which limits the usefulness of these cultures. This decreased functionality could be the result of adaptive responses of hepatocytes to the new culture environment or to the lack of elements in the culture media that are essential to the in vivo hepatocellular function (Suolinna 1982; Holme 1985; Guillouzo 1986; Falasca et al. 1998). Hepatic in vitro models that retain hepatocellular structure and express differentiated metabolic functions are therefore needed for long-term studies on hepatic functionality, enzymatic induction, and drug-drug interactions.

In the intact liver, hepatocytes are in close three-dimensional contact with each other, and extracellular matrix proteins present in the Disse spaces also attach hepatocytes (Rojkind 1982; Martinez-Hernández 1984). Therefore, the maintenance of the differentiated status, as is found in the adult liver, may require not only soluble signals, which can be released in culture media, but also cell-cell and cell-matrix interactions, which are difficult to mimic in conventional plastic-cultured cells but maintain cell structure and regulate the transcription of hepatic genes (Caron 1990; Dunn et al. 1992; Nagaki et al. 1995; Oda et al. 1995; Gómez-Lechón et al. 1998).

Several strategies for mimicking the microenvironment of the liver in vitro have been explored: supplementation of the culture medium with hormones and cofactors; co-culture of hepatocytes with nonparenchymal cells; and the use of extracellular matrix proteins (Fig. 1). The use of a hormonally defined medium can partially prevent the loss of liver-specific functions, but does not promote the survival of hepatocytes for extended periods of time. Cell-cell contacts between hepatic cells seem to be important in the expression of the cellular functions of hepatocytes in the liver (Rojkind 1982; Goulet et al. 1988). One way to reproduce this microenvironment in vitro is to bring hepatocytes

[1] Unidad de Hepatología Experimental, Centro de Investigación, Hospital La Fe. Avda. Campanar 21, 46009 Valencia, Spain
[2] Biología Celular, Facultad Biológicas. Avda. Dr. Moliner 50. 46100 Burjasot, Spain

Progress in Molecular and Subcellular Biology, Vol. 25
A. Macieira-Coelho (Ed.)
© Springer-Verlag Berlin Heidelberg 2000

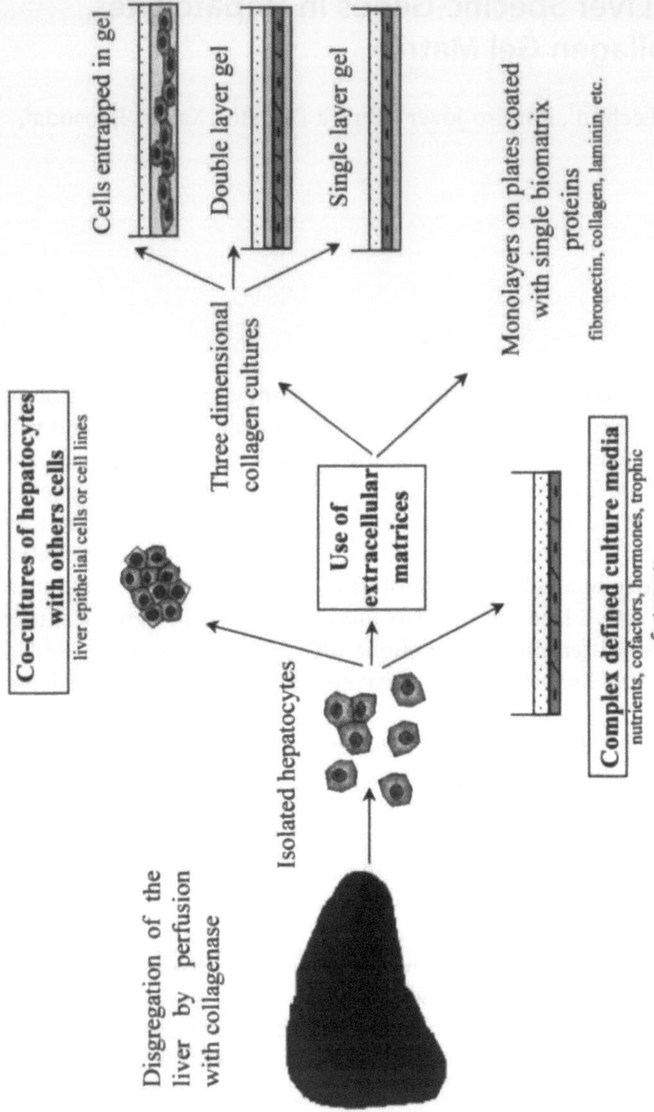

Fig. 1. Strategies for improving and prolonging the expression of hepatic phenotype in cultured hepatocytes

into close contact with nonparenchymal cells by co-culturing them. This idea, first described by Langenbach et al. (1979) and Michalopoulos et al. (1979) using fibroblasts as feeder-layers, was later improved by Guguen-Guillouzo et al. (1983), who reported that by co-culturing rat hepatocytes with rat liver epithelial cells, the hepatic adult phenotype was expressed for a much longer time than in conventional pure cultures. The nature of the helping cells does not seem to be absolutely critical for their positive role in co-cultures (Morin and Norman 1986; Goulet et al. 1988; Donato et al. 1992a) and, in fact, established cell lines of different origin have been used uccessfully in co-culture with rat hepatocytes (Morin and Normand 1986; Kuri-Harcuch and Mendoza-Figueroa 1989; Donato et al. 1990, 1991, 1994). Co-cultures allow the long-term culture of hepatocytes where a close contact between both types of cells stimulates hepatocytes to retain their differentiated adult phenotype longer than conventional pure cultures (Langenbach et al. 1979; Michalopoulos et al. 1979; Begué et al. 1984; Rogiers et al. 1990; Donato et al. 1991, 1992a, b, 1994; Lerche et al. 1997; Tateno and Yoshizato 1998).

The key role of the extracellular matrix structures including laminin, fibronectin, collagen, and glycosaminoglycans, in cell attachment, morphogenesis, and in the regulation of liver gene expression was recognized years ago in rat hepatocyte cultures (Michalopoulos and Pitot 1975; Michalopoulos et al. 1976; Rojkind et al. 1980; Reid et al. 1986; Dunn et al. 1989, 1992; Elcin et al. 1998), and attempts have been made to reproduce these complex cell-biomatrix interactions in culture. The use of the liver-derived biomatrix was tested, but its poorly defined composition and the variability between different preparations prevented its generalized use (Rojkind et al. 1980; Reid et al. 1986). Single proteins from the extracellular matrix, such as laminin (Bissell et al. 1987) and fibronectin (Yamada 1982; Gómez-Lechón et al. 1992) have been used, as have other extracellular matrix components (te Velde et al. 1995) and Matrigel (Guzelian et al. 1988; Brown et al. 1995; Ikeda et al. 1998; Schmoelzl et al. 1998), a solubilized preparation of basement membrane components derived from the mouse Engelbreth-Holm-Sarcoma (EHS) (Guzelian et al. 1988). Evidence is now accumulating of the positive role of the extracellular matrix in the maintenance of hepatocyte polarity (Le Cluyse et al. 1994), biochemical activity, and differentiated functions of hepatocytes in primary culture (Dunn et al. 1992; Koebe et al. 1994; Bader et al. 1994, 1996; Brown et al. 1995; Nakajima and Shimbara, 1996; Gómez-Lechón et al. 1998; Liu et al. 1998; Santhosh et al. 1998). However, organotypic cultures in which hepatocytes contact three-dimensionally with extracellular matrix proteins and do not simply coat the plates have been shown to be the most effective for improving the hepatocellular functions (Koebe et al. 1994; Bader et al. 1996; Gómez-Lechón et al. 1998; Kono et al. 1998; Liu et al. 1998) and maintaining hepatocyte in vivo structure, polarity, and bile canaliculi development (Le Cluyse et al. 1994; Gómez-Lechón et al. 1998). In addition, by using confocal laser-scanning microscopy, the three-dimensional organization of cytoplasmic microtubules of hepatocytes cultured in Matrigel and collagen gels has been analyzed, and these studies suggest that the organization of cytoplasmic microtubules participates in the shape-related regulation of cell function (Ikeda et al. 1998). In this context, collagen gels represent a simplified alternative to more complex biomatrices that consists of setting hepatocytes either in a sandwich configuration between two layers of collagen (Dunn et al. 1989, 1992; Bader et al. 1994, 1996) or entrapping hepatocytes within a homogeneous collagen gel (Koebe et al. 1994; Gómez-Lechón et al. 1998). By culturing cells in this fashion the situation in the liver (two-side exposure of hepatocytes within the Disse space) can be mimicked (Fig. 1). This progress has encouraged the use of three-dimensional culture systems in bioreactors and in bioartificial liver support (Nyberg et al. 1992; Sato et al. 1994; Lin et al. 1995; Dixit and Gitnick 1996; Lopina et al. 1996; Jowiak et al. 1998).

Among the above-mentioned systems, culture in collagen gels is a reproducible, technically feasible, and efficient way to reproduce the in vivo three-dimensional microenvironment of hepatocytes.

2
Culture of Hepatocytes in Collagen Gel

Hepatocytes are isolated from Sprague-Dawley male rats (180–250g) by reverse perfusion of the liver with collagenase, as previously described by Gómez-Lechón et al. (1992), and are resuspended in Ham F-12/Leibovitz −15 medium supplemented with 2% newborn calf serum, 30 mM proline (Lee et al. 1993), 0.2% BSA, 10^{-8} M insulin and antibiotics. Finally, hepatocytes are seeded either on fibronectin ($3.5 \, \mu g \, cm^{-2}$) coated culture plates or entrapped in collagen type-I at a final density of 80×10^3 and 150×10^3 viable cells cm-² respectively. Collagen is prepared immediately before use by mixing 9 vol. of a $1.5 \, mg \, ml^{-1}$ collagen solution in 1 mM HCl (Koebe et al. 1994) and 1 vol. of 10× concentrated culture medium. Hepatocytes are gently dispersed 1:1, v/v with collagen solution. The cell suspension is poured onto plates and allowed to gelify. Fresh culture medium is finally added to plates. The cultures are shifted to serum-free hormone-supplemented medium (10^{-8} M of dexamethasone and insulin) 24h later. Thereafter medium is renewed daily. Hepatic functions, biotransformation activities, and transcription factor expression are determined as previously described in detail (López et al. 1984; Castell et al. 1985; Gómez-Lechón et al. 1996, 1998).

3
Hepatic Specific Functions of Hepatocytes in Collagen Cultures

Hepatocyte gene expression is known to be dependent on extracellular matrix composition, and cells cultured in collagen gels survive for long terms, and maintain their capability to respond to hormones and the expression of the differentiated hepatocyte phenotype (Dunn et al. 1992; Koebe et al. 1994; Bader et al. 1994, 1996; Brown et al., 1995; Nakajima and Shimbara 1996; Gómez-Lechón et al. 1998; Liu et al. 1998; Santhosh et al. 1998).

A major function of the liver in homeostasis in the body is the control of the glucose concentration in blood during the postabsorptive state and fasting. This involves the activation of synthesis and mobilization of glycogen and gluconeogenesis. The maintenance of these processes, which are known to be dependent on nutritional and hormonal factors, has been investigated in hepatocytes entrapped in collagen. Hepatocytes respond to glucagon by mobilizing ca. 80% of their glycogen stores. Synthesis and accumulation of glycogen by glycogen-depleted cells are stimulated by physiological concentrations of insulin, which leads to increased intracellular glycogen content, similar to that of liver and hepatocytes isolated from fed rats (Fig. 2A). Hepatocytes are able to synthesize glucose from lactate, the physiological precursor, for extended periods in culture (Gómez-Lechón et al. 1998).

Regarding lipid metabolism, it has recently been reported that hepatocyte lipoprotein receptor expression is also better maintained at the mRNA, protein and functional levels in hepatocytes cultured on Matrigel and in collagen type I (Schmoelzl et al. 1998).

Two other typical hepatic functions are urea and plasma protein synthesis. The synthesis and secretion of urea and albumin to the external medium by hepatocytes was also analyzed. Maximal capacity for urea synthesis after overloading cells with ammonia remained stable until the end of culture (Fig. 2B). Albumin synthesis increased during the first 7 days of culture and then gradually decreased (Fig. 2C), as reported by other authors (Nakajima and Shimbara 1996; Kono et al. 1998).

4
Biotransformation Capability:
Basal and Induced Levels of Drug-Metabolizing Enzymes

One of the major drawbacks of hepatocytes cultured on plastic is the rapid loss of CYP enzymes and the limited ability of cultured cells to respond to enzyme inducers (Donato et al. 1992b). Hepatocytes entrapped in collagen gel express phase I and phase II enzymes for more than 2 weeks (Gómez-Lechón et al. 1998; Silva et al. 1998).

The activity of cytochrome P450 isozymes (CYP) is clearly measurable after 18 days of culture. Basal levels of 7-ethoxyresorufin O-deethylase (EROD, CYP1A1/2) and 7-pentoxyresorufin O-depentylase (PROD, CYP2B1) activities (Fig. 3A) are maintained fairly unchanged for several days in culture. Testosterone hydroxylation at the 6β position activity (CYP3A1; Fig. 3B), chlorzoxazone 6-hydroxylation (CYP2E1; Fig. 3B) and diclofenac 4'-hydroxylation (CYP2C11; Fig. 3B) show an initial decrease, but maintain quite stable levels thereafter.

After 4 days in culture, the conjugation with glutathione by glutathione transferase (GST), a critical pathway in cell self-protection, recovers to the

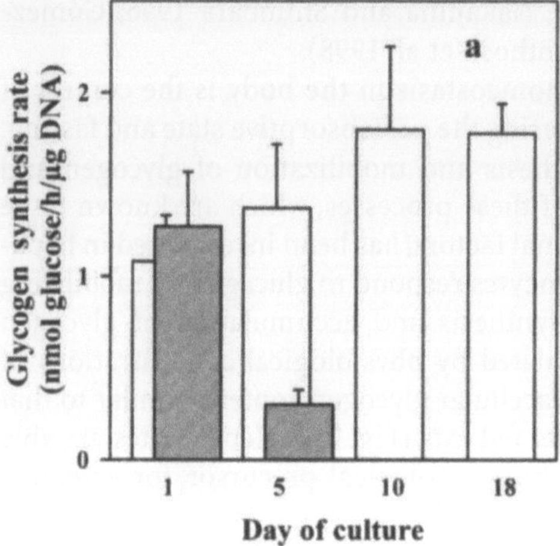

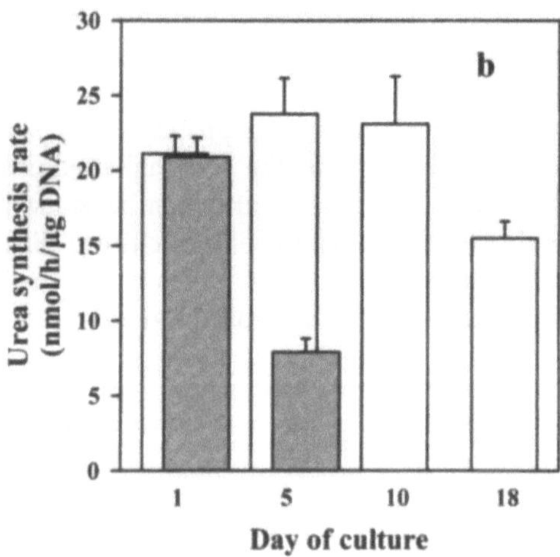

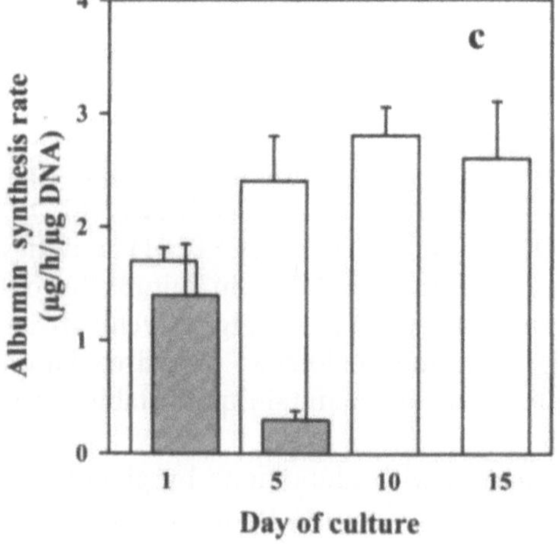

Fig. 2a–c. Hepatic metabolic functions in hepatocytes immobilized in collagen gel. Glycogen synthesis rate in response to insulin (**a**); urea synthesis rate (**b**); and albumin synthesis rate (**c**) were measured in collagen cultures (*open bars*) and in plastic cultures (*shaded bars*). Synthesis and accumulation of glycogen was measured in glycogen-depleted cells by 2-h pretreatment with 10^{-7} M glucagon and further stimulation with 10^{-8} M insulin and 30 mM glucose. Rate of urea synthesis was determined in cultures overloaded with 4 mM ammonia for 2h. The rate of albumin secretion was determined by ELISA in aliquots of culture medium

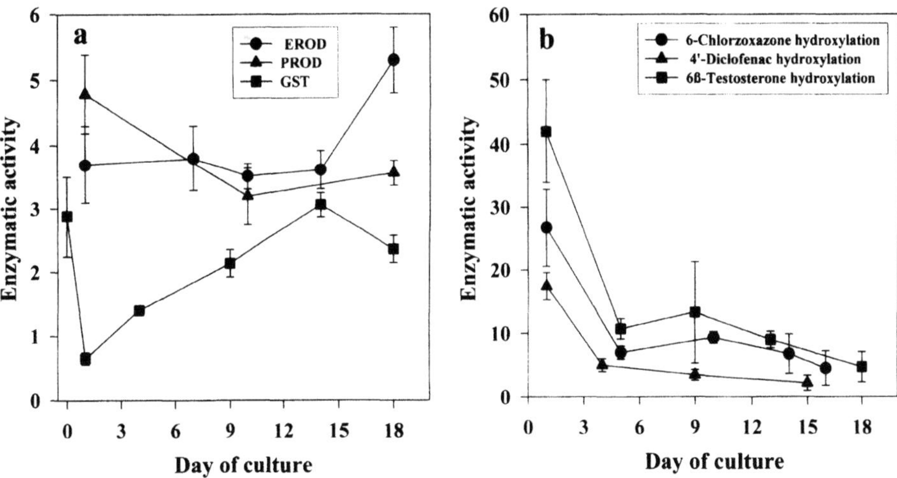

Fig. 3a,b. Drug-metabolizing activities in collagen-entrapped hepatocytes. Timecourse of basal metabolic rates of CYP2A1/2 (EROD activity), CYP2B1 (PROD activity) and GST activity (**a**), and CYP2E1 (chlorzoxazone hydroxylation), CYP2C11 (4'-diclofenac hydroxylation) and CYP3A1 (testosterone 6β hydroxylation) (**b**) in hepatocytes entrapped on collagen gel. EROD, PROD, 6-chlorzoxazone hydroxylation, 4'-diclofenac hydroxylation and testosterone 6β hydroxylation activities are expressed as $pmol\,h^{-1}\,mg^{-1}$ cellular protein. GST activity is expressed as $\mu mol\,h^{-1}\,mg^{-1}$ cellular protein

levels of isolated hepatocytes and remains fairly constant thereafter (Fig. 3A), while on plastic cultured hepatocytes it gradually decreases, as reported by Donato et al. (1992b). The operativeness of this pathway is ensured by sustained high levels of the tripeptide GSH culture ($1295 \pm 118\,pmol\,\mu g^{-1}$ DNA).

Induction of CYPs by drugs can lead to drug-drug interactions that may have relevant clinical pharmacological and toxicological consequences for the elimination and the therapeutic efficacy of drugs. In vitro investigation of these processes should answer the demand for a fast and reproducible method for addressing CYP induction/inhibition by numerous compounds at the drug discovery stage. Therefore, the optimization of culture conditions by using extracellular matrixes may allow resarchers to use cultured hepatocytes for predicting induction of CYP expression in vivo.

The loss of CYP expression in conventional cultures in plastic is due to cessation of the transcriptional activity and/or changes in specific mRNAs or protein stability.

The long-term inducibility of two CYPs, CYPA1/2 and CYP2E1, whose activity is increased in hepatocytes by one of the two above-mentioned mechanisms, has been studied. The inducibility of both CYPs in collagen gel-entrapped hepatocytes was consistently higher than in hepatocytes cultured in plastic, as has also been reported by other authors (Silva et al. 1998). Treatment

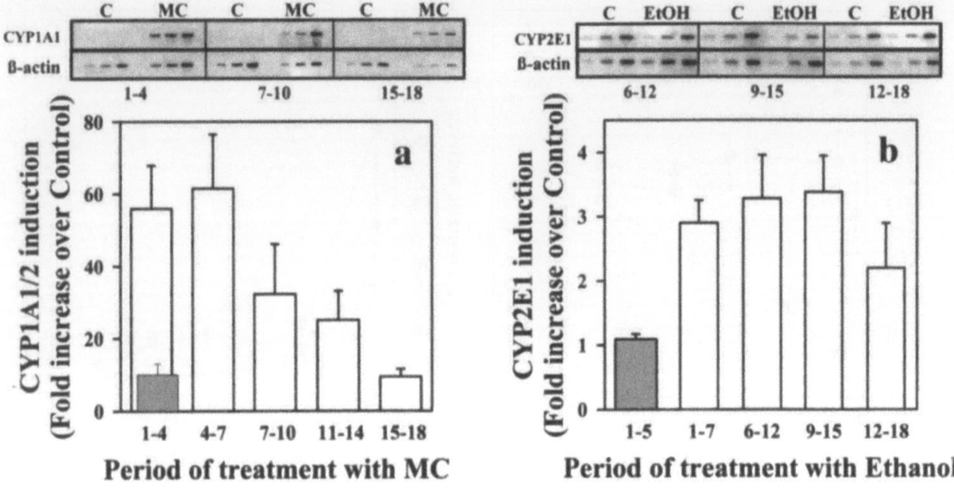

Fig. 4a,b. Induction of activity and changes in mRNA levels of CYP1A2 and CYP2E1 in hepatocytes exposed to methylcholanthene and ethanol. Hepatocytes were treated with 2 µM methylcholanthrene (a) or 75 mM ethanol (b) for 3 and 6 days, respectively, starting on a different day of culture. Data are expressed as the fold variation of EROD and chorzoxazone hydroxylation activities with respect to the control Collagen cultures (*open bars*) and plastic cultures (*hatched bars*). Changes in CYP1A1 and CYP2E1 mRNA levels in methylcholanthrene and ethanoltreated cells and in controls were determined by semiquantitative RT-PCR

with methylcholanthrene (MC) resulted in a 60–90-fold increase in EROD activity over control (Fig. 4A), while chlorzoxazone hydroxylation after exposure to ethanol increased more moderately (4–6-fold over control; Fig. 4B). The results make it clear that collagen-entrapped hepatocytes could respond to enzyme inducers despite long periods in culture (Fig. 4), which is consistent with the findings of Nakajima and Shimbara (1996).

The induction of CYP1A1 and CYP1A2 by MC was fast (2 or 3 days of exposure to MC), and the enzyme activity was paralleled by increased levels of the specific mRNA, thus indicating that the transcriptional activation of the CYP1A1 and CYP1A2 genes by MC is a conserved mechanism (Okey 1990). Exposure of hepatocytes to ethanol significantly induced CYP2E1 activity without concomitant changes in the corresponding mRNA, as has been reported in cultured hepatocytes as well as *in vivo* (Eliasson et al. 1988; Kim et al. 1995; Roberts et al. 1995). This suggests that the increased enzyme activity measured in hepatocytes presumably occurs at the translational or postranslational level, and apparently excludes increased levels of transcription or mRNA stabilization as major mechanisms.

5
Expression of Hepatic Transcription Factors

Serum-free, hormonally defined conditions and extracellular matrix requirements were investigated for cell differentiation and regulation of hepatic-
specific gene expression. Culturing hepatocytes in collagen matrix has a positive effects on the stable maintenance of the expression of other hepatic-
specific genes, in addition to those mentioned previously. Analysis of albumin,
phosphoenol pyruvate carboxykinase, carbamylphosphate syntethase, CYPs,
and tyrosin aminotransferase mRNAs revealed consistently higher levels of
expression in collagen cultures than in the same cells cultured in plastic dishes,
and this points to a more efficient gene transcription in collagen cultures
(Gómez-Lechón et al. 1998).

The differential activation of hepatic gene expression in the adult hepatocyte is controlled by the so-called liver-enriched transcription factors
(Cereghini 1996). However, an early, and fast decay has been reported in the
expression of most of them during hepatocyte isolation and the initial stages
of conventional cultures, which parallels a decrease in hepatic-specific gene
expression (Padham and Paine 1993). Moreover, hepatic transcription factors,
particularly C/EBP-α and HNF-3, show very low levels in dedifferentiated
hepatoma cell lines (Jover et al. 1998). In contrast, a direct relationship between
differentiation of hepatocytes and hepatoma cells and expression of transcription factors, in which the hepatic factors HNF-4 and C/EBP-α play a major
role, has been described (Griffo et al. 1993; Spath and Weiss 1997; Jover et al.
1998; Runge et al. 1998). Therefore, the expression of the most relevant
families of liver-enriched transcription factors (C/EBP-α, PAR, HNF-1, 3 and
4) was analyzed in hepatocytes cultured in collagen gels and compared with
conventional cultures.

In collagen cultures the hepatic nuclear factors HNF-1 and 4, as well as
C/EBP-α maintained significantly higher mRNA levels than with day-1 hepatocytes cultured in plastic. Interestingly, HNF1 and particularly HNF-4 (12-fold
increase over day 1) are not only maintained but greatly increase in collagen
cultures (Fig. 5). These findings suggest that there is a concomitant increase
in the expression of target genes (such as albumin, phosphoenol pyruvate
carboxykinase, carbamylphosphate syntethase, CYPs, and tyrosin aminotransferase) and the levels of hepatic transcription factors (C/EBP-α, HNF1, 3
and 4) which are directly involved in hepatic gene regulation.

C/EBP-α, which is detected in abundance in differentiated hepatocytes,
transactivates the promoters of several hepatocyte specific genes including
albumin (Friedman et al. 1989), phosphoenol pyruvate carboxykinase
(Croniger et al. 1998), and CYPs (Phuong-Van et al. 1996; Ourlin et al. 1997),
and acts in conjunction with other transcription factors (NHF-1, DBP and
C/EBP-β). HNF-4 seems to play a key role as a major common regulator for the
liver-specific transcriptional activation of many CYP genes, i.e., CYP3A1/2
(Yokomori et al. 1997; Huss and Kasper 1998; Ogino et al. 1999), CYP2C (Chen

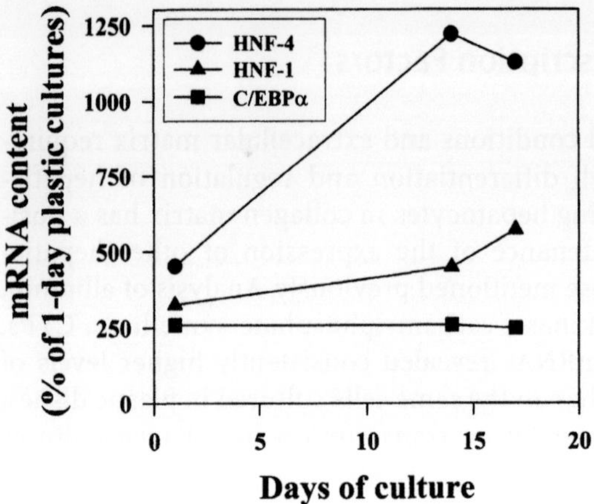

Days of culture

Fig. 5. Expression of hepatic transcription factors in hepatocytes entrapped in collagen gel. Changes in HNF1 and HNF-4 and C/EBP-α mRNA were determined by semiquantitative RT-PCR in hepatocytes cultured for 1 day in plastic culture dishes and from hepatocytes cultured for 1, 14, and 17 days in collagen. Data are expressed as a percentage of mRNA content measured in 1-day-cultured hepatocytes in plastic

et al. 1994; Legraverend et al. 1994), CYP2D6 (Cairns et al. 1996), and albumin (Griffo et al. 1993; Runge et al. 1998), and for the activation of factor HNF-1 (Cereghini 1996). HNF-1 has also been reported to transactivate CYP2E1 (Liu and Gonzalez 1995) and albumin (Rollier et al. 1993).

The proper expression and transactivation of hepatic genes usually depends on a transcriptional network in which the synergistic and cooperative interactions of two or more transcription factors operate (Herbomet 1990). For instance, the ornithine transcarbamylase enhancer requires both HNF-4 and C/EBP-β for functional activation (Nishiyori et al. 1994), and HNF-4 activates HNF-1 (Cereghini 1996). Therefore, as shown in Fig. 5, the simultaneous expression of sustained high levels of the most relevant hepatic transcription factors (C/EBP-α, HNF-1 and in particular HNF-4) is the key to an efficient, extended expression of hepatic genes in collagen cultures (Oda et al. 1995; Späth and Weiss 1997, 1998).

Differentiation of cells is governed by transcription factors, but these seem to be under the control of the combinatorial effect of hormonal and cell-cell and cell-matrix interactions. By integrating these three clues, the transcription factors may synergistically function by promoting cell differentiation (Smas and Sul 1997). The extracellular matrix components regulate cell function by interacting with specific members of the integrin family of cell surface receptors (Boudreau and Bissell 1998; Longhurst and Jennings 1998; Dedhar 1999). Integrins transduce internal signals from the extracellular matrix by binding through their extracellular domain to a specific peptide recognition site ligand, and signals generated from ligand-integrin interactions are propagated via the integrin cytoplasmic tail to signal transduction pathways within the cell to the nucleus (Boudreau and Bissell 1998; Longhurst and Jennings 1998; Dedhar 1999). It has recently been reported that by mediating interactions between cells and the extracellular matrix, integrins play an important role in the differentiation of the epithelial and endothelial cell populations during human liver organogenesis (Couvelard et al. 1998). Therefore, it could be hypothesized that collagen initiates a transmembrane signaling pathway in hepatocytes that activates specific hepatic gene expression and cell differentiation.

6
Concluding Remarks

The fact that the liver-specific transcription factors are expressed better and longer in collagen than in plastic cultures strongly suggests that collagen could stimulate cell differentiation and long-term expression of hepatocyte genes by activating liver-specific transcription factors. Maintaining the differentiated status of hepatocytes may require not only soluble signals (that can be delivered in culture media) but also cell-matrix interactions, which are difficult to mimic in conventional plastic-cultured cells but regulate liver-specific transcription factors (Caron 1990; Dunn et al. 1992; Nagaki et al. 1995; Oda et al. 1995). This is critical when the induction of hepatic genes is to be studied after several days in culture (e.g., CYP genes).

It can be concluded that immobilization of hepatocytes within a three-dimensional matrix of collagen type I shows a positive influence on the survival, well-preserved morphology and functionality, as well as a high expression of both constitutive and inducible hepatic genes in comparison with the same cell preparations cultured on plastic.

Acknowledgments. The financial support of European Union and the ALIVE Foundation are gratefully acknowledged.

References

Bader A, Zech K, Crome O, Christians U, Ringe B, Pichlmayer R, Sewing K FR (1994) Use of organotypical cultures of primary hepatocytes to analyse drug biotransformationm in man and animals. Xenobiotica 2:623–633

Bader A, Knop E, Kern A, Böker K, Frühauf N, Cromee O, Esselmann H, Pape C, Kempa G, Sewing K FR (1996) 3-D Coculture of sinusoidal dells with primary hepatocytes. Design of an organotypical model. Exp Cell Res 226:223–233

Begué J, Guguen-Guillouzo C, Pasdeloup N, Guillouzo A (1984) Prolonged maintenance of active cytochrome P-450 in adult rat hepatocyte co-cultured with another liver cell type. Hepatology 4:839–842

Bissell DM, Arenson DM, Maher JJ, Roll FJ (1987) Support of cultured hepatocytes by a laminin-rich gel. Evidence for a functionally significant subendothelial matrix in normal rat liver. J Clin Invest 7:801–812

Boudreau N, Bissell MJ (1998) Extracellular matrix signaling: integration of form and function in normal and malignant cells. Curr Opin Cell Biol 10:640–646

Brown SE, Guzelian ChP, Schuetz E, Quattrochi LC, Kleinman HK, Guzelian PS (1995) Critical role of the extracellular matrix on induction by phenobarbital of cytochrome P450 2B1/2 in primary cultures of adult rat hepatocytes. Lab Invest 6:818–827

Caron JM (1990) Induction of albumin gene transcription in hepatocytes by extracellular matrix proteins. Mol Cell Biol 10:1239–1243

Cairns W, Smith CAD, McLaren AW, Wolf CR (1996) Characterization of the human cytochrome P4502D6 promoter. A potential role for antagonistic interactions between members of the nuclear receptor family. J Biol Chem 271:25269–25276

Castell JV, Larrauri A, Gómez-Lechón MJ (1985) A study of the relative hepatotoxicity in vitro of the non-steroidal anti-inflammatory drugs ibuprofen, flurbiprofen and butibufen. Xenobiotica 18:737–745

Castell JV, Bort R, Ponsoda X, Gómez-Lechón MJ (1997) In vitro investigation of the molecular mechanisms of hepatotoxicity. In: Castell JV, Gómez-Lechón MJ (eds) In vitro methods in pharmaceutical research. Academic Press, London, pp 375–432

Cereghini S (1996) Liver-enriched transcription factors and hepatocyte differentiation. FASEB J 10:267–282

Couvelard A, Bringuier AF, Dauge MC, Nejjari M, Darai E, Benifla JL, Feldman G, Henin D, Scoazec JY (1998) Expression of integrins during liver organogenesis in humans. Hepatology 27:839–847

Croniger C, Leahy P, Reshef L, Hanson RW (1998) C/EBP and control of phosphoenolpyruvate carboxykinase gene transcription in the liver. J Biol Chem 273:31629–31633

Chen D, Park Y, Kemper B (1994) Differential protein binding and transcriptional activities of HNF-4 elements in three closely related CYP2C genes. DNA Cell Biol 13:771–779

Dedhar S (1999) Integrins and signal transduction. Curr Opin Hematol 6:37–43

Dixit V, Gitnick G (1996) Artificial liver support: state of the art. Scand J Gastroenterol Suppl 220:101–104

Donato MT, Gómez-Lechón MJ, Castell JV (1990) Drug metabolizing enzymes in rat hepatocytes co-cultured with cell lines. In Vitro Cell Dev Biol 26:1057–1063

Donato MT, Castell JV, Gómez-Lechón MJ (1991) Co-cultures of hepatocytes with epithelial-like cell lines: expression of drug-biotransformation activities by hepatocytes. Cell Biol Toxicol 7:1–14

Donato MT, Gómez-Lechón MJ, Castell JV (1992a) Co-cultures of hepatocytes: A biological model for long-lasting cultures. In: Castell JV, Gómez-Lechón MJ (eds) In vitro alternatives to animal pharmacotoxicology. Farmaindustria, Barcelona, pp 109–127

Donato MT, Gómez-Lechón MJ, Castell JV (1992b) Biotransformation of drugs by cultured hepatocytes. In: Castell JV, Gómez-Lechón MJ (eds) In vitro alternatives to animal pharmacotoxicology. Farmaindustria, Barcelona, pp 149–178

Donato MT, Castell JV, Gómez-Lechón MJ (1994) Cytochrome P450 activities in pure and co-cultured rat hepatocytes. Effects of model inducers. In Vitro Cell Dev Biol 30A:825–832

Dunn JCY, Yarmush MKL, Koebe HG, Tomkins RG (1989) Hepatocyte funtion and extracellular matrix geometry: long-term culture in a sandwich configuration FASEB J 3:174–177

Dunn JCY, Tompkins RG, Yarmush ML (1992) Hepatocytes in collagen sandwich: evidence for transcriptional and translational regulation J Cell Biol 116:1043–1053

Elcin YM, Dixit V, Gitnick G (1998) Hepatocyte attachment on biodegradable modified chitosan membranes: in vitro evaluation for the development of liver organoids. Artif-Organs 22:837–846

Eliasson E, Johansson I, Ingelman-Sundberg M (1988) Ligand-dependent maintenance of ethanol-inducible cytochrome P-450 in primary rat hepatocyte cell cultures. Biochem Biophys Res Commun 150:436–443

Falasca L, Favale A, Serafino A, Ara C, Conti-Devirgiliis L (1998) The effect of retinoic acid on the re-establishment of differentiated hepatocyte phenotype in primary culture. Cell Tissue Res 293:337–347

Friedman AD, Landschulz WH, McKnight SL (1989) CCAAT/enhancer binding protein activates the promoter of the serum albumin in cultured hepatoma cells. Gene Dev 3:1314–1322

Gómez-Lechón MJ, López P, Castell JV (1984) Biochemical functionality and recovery of hepatocytes after deep freezing storage. In Vitro 20:826–832

Gómez-Lechón MJ, Donato MT, Ponsoda X, Jover R, Castell JV (1992) Experimental in vitro models to evaluate hepatotoxicity. In: Castell J, Gómez-Lechón MJ (eds) In vitro alternatives to animal pharmacotoxicology, Farmaindustria, Barcelona, pp 85–92

Gómez-Lechón MJ, Ponsoda X, Castell JV (1996) A microassay for measuring glycogen in 96-well-cultured cells. Anal Biochem 263:296–301

Gómez-Lechón MJ, Jover R, Donato MT, Ponsoda X, Rodriguez C, Stenzel KG, Klocke R, Paul D, Guillen I, Bort R, Castell JV (1998) Long-term expression of differentiated functions in hepatocytes cultured in three-dimensional collagen matrix. J Cell Physiol 177:553–562

Goulet F, Normand C, Morin O (1988) Cellular interactions promote tissue-specific function, biomatrix deposition and junctional communication of primary cultured hepatocytes. Hepatology 8:1010–1018

Griffo G, Hamon-Benais C, Angrand PO, Fox M, West-L, Lecoq-O, Povey S, Cassio D, Weiss M (1993) HNF4 and HNF1 as well as a panel of hepatic functions are extinguished and reexpressed in parallel in chromosomally reduced rat hepatoma-human fibroblast hybrids. J Cell Biol 121:887–898

Guguen-Guillouzo Ch, Clement B, Baffet G, Beaumont C, Morel-Chany E, Glaise D, Guillouzo A (1983) Maintenance and reversibility of active albumin secretion y adult rat hepatocytes cocultured with another liver epithelial cell type. Hepatology 4:839–842

Guillouzo A (1986) Use of isolated and cultured hepatocytes for xenobiotic metabolism and cytotoxicity studies. In: Guillouzo A, Guguen-Guillouzo Ch (eds) Research in isolated and culture hepatocytes, INSERM/John Libbey Eurotext, London, pp 313–332

Guzelian PS, Li D, Schuetz EG, Thomas P, Levin W, Mode A, Gustafsson JA (1988) Sex change in cytochrome P-450 phenotype by growth hormone treatment of adult rat hepatocytes maintained in a culture system on matrigel. Proc Natln Acad Sci USA 85:9783–9788

Herbomet P (1990) Synergistic activation of eukaryotic transcription: the multiacceptor target hypothesis. New Biol 2:1063–1070

Holme JA (1985) Isolation and culture of liver cells and their use in the biochemical research of xenobiotics. NIPH Ann 8:49–63

Huss JM, Kasper CB (1998) Nuclear receptor involvement in the regulation of rat cytochrome P450 3A23 expression. J Biol Chem 273:16155–16162

Ikeda T, Sawada N, Satoh M, Mori M (1998) Induction of tyrosine aminotransferase of primary cultured rat hepatocytes depends on the organization of microtubules. J Cell Physiol 175:41–49

Jover R, Bort R, Gómez-Lechón MJ, Castell JV (1998) Re-expression of C/EBP-α c induces CYP2B6, CYP2C9 and CYP2D6 genes in 2 in HepG2 cells. FEBS Lett 431:227–230

Jozwiak A, Karlik W, Wiechetek M, Werynski A (1998) Attachment and metabolic activity of hepatocytes cultivated on selected polymeric membranes. Int J Artif Organs 21:460–466

Kim SG, Shehin SE, States Ch, Novak RF (1995) Evidence for increased translational efficiency in the induction of P450IIE1 by solvents: analysis of P450IIE1 mRNA polyribosomal distribution. Biochem Biophys Res Commun 172:767–774

Koebe HG, Pahernik S, Eyer P, Schildberg FW (1994) Collagen gell immobilization: a usefull cell culture technique for long-term metabolic studies on human hepatocytes. Xenobiotica 24:95–107

Kono K, Yang S, Roberts EA (1998) Extended primary culture of hunman hepatocytes in a collagen gel sandwich system. In Vitro Cell Dev Biol Animal 33:467–472

Kuri-Harcuch W, Mendoza-Figueroa T (1989) Cultivation of adult rat hepatocytes on 3T3 cells: expression of various liver differentiated functions. Differentiation 41:148–157

Langenbach R, Malock L, Tompa A, Kuszynski C, Freed H, Huberman E (1979) Maintenance of adult rat hepatocytes on C3H/10T1/2 cells. Cancer Res 39:3509–3514

LeCluyse EL, Audus KL, Hochman JH (1994) Formation of extensive canalicular networks by rat hepatocytes cultured in collagen-sandwich configuration. Am J Physiol 266:C1764–C1774

Lee J, Morgan JRM, Tompkins RG, Yarmush ML (1993) Proline-mediated enhacement of hepatocyte function in a collagen gel sandwich culture configuration. FASEB J 7:586–591

Legraverend C, Eguchi H, Strom A, Lahuna O, Mode A, Tollet P, Westin S, Gustafsson JA (1994) Transactivation of the rat CYP2C13 gene promoter involves HNF-1, HNF-3, and members of the orphan receptor subfamily. Biochemistry 33:9889–9897

Lerche C, Fautrel A, Shaw PM, Glaise D, Ballet F, Guillouzo A, Corcos L (1997) Regulation of the major detoxication functions by phenobarbital and 3-methylcholanthrene in co-cultures of rat hepatocytes and liver epithelial cells. Eur J Biochem 244:98–106

Lin KH, Maeda S, Saito T (1995) Long-term maintenance of liver-specific functions in three-dimensional culture of adult rat hepatocytes with porous gelatin sponge support. Biotechnol Appl Biochem 21:19–27

Liu SY, Gonzalez FJ (1995) Role of the liver enriched transcription factor HNF-1α in expression of CYP2E1 gene. DNA Cell Biol 4:285–293

Liu X, Brouwer KL, Gan LS, Brouwer KR, Stieger B, Meier PJ, Audus KL, LeCluyse EL (1998) Partial maintenance of taurocholate uptake by adult rat hepatocytes cultured in a collagen sandwich configuration. Pharm Res 15:1533–1539

Longhurst CM, Jennings LK (1998) Integrin-mediated transduction. Cell Mol Life Sci 54:514–526

López MP, Gómez-Lechón MJ, Castell JV (1984) Glycogen synthesis in serum-free cultured hepatocytes in response to insulin and dexamethasone. In Vitro 20:923–931

Lopina ST, Wu G, Merrill EW, Griffith-Cima L (1996) Hepatocyte culture on carbohydrate-modified star polyethylene oxide hydrogels. Biomaterials 17:559–569

Maher JJ (1988) Primary hepatocyte culture: is it home away from home? Hepatology 8:1162–1166

Martinez-Hernández A (1984) The hepatic extracellular matrix. I. Electron immunohistochemical studies in normal rat liver. Lab Invest 51:57–74

Michalopoulos G, Pitot HC (1975) Primary culture of parenchymal liver cells on collagen membranes. Exp Cell Res 94:70–78

Michalopoulos G, Sattler GL, Pitot HC (1976) Maintenance of microsomal cytochrome b5 and P-450 in primary cultures of parenchymal liver cells on collagen membranes. Life Sci 18:1139–1144

Michalopoulos G, Russell F, Biles C (1979) Primary cultures of hepatocytes on human fibroblasts. In Vitro 15:796–806

Morin O, Normand C (1986) Long-Term maintenance of hepatocyte functional activity in co-culture: requeriments for sinusoidal endothelial cells and dexamethasone. J Cell Physiol 129:103–110

Nagaki M, Shidoji Y, Yamada Y, Sugiyama A, Tanaka M, Akaike T, Ohishi H, Moriwaki H, Muto Y (1995) Regulation of hepatic genes and liver transcription factors in rat hepatocytes by extracellular matrix. Biochem Biophys Res Commun 210:38–43

Nakajima H, Shimbara N (1996) Functional mainance of hepatocytes on collagen gel cultured with simple serum-free medium containing sodium selenite. Biochem Biophys Res Commun 222:664–668

Nishiyori A, Tashiro H, Kimura A, Akagi K, Yamamura K, Mori M, Takiguchi M (1994) Determination of tissue specificity of the enhancer by combinatorial operation of tissue-enriched transcription factors. Both HNF-4 and C/EBP-β are required for liver-specific activity of the ornithine transcarbamylase enhancer. J Biol Chem 269:1323–1333

Nyberg SL, Shatford RA, Payne WD, Hu W, Cerra FB (1992) Primary culture of rat hepatocytes entrapped in cylindrical collagen gels: an in vitro system with application to the bioartificial liver. Cytotechnology 10:205–215

Oda H, Nozawa K, Hitomo Y, Kakinuma A (1995) Laminin-rich extracellular matrix maintains high levels of hepatocyte nuclear factor 4 in rat hepatocyte culture. Biochem Biophys Res Commun 212:800–805

Ogino M, Nagata K, Miyata M, Yamazoe Y (1999) Hepatocyte nuclear factor 4-mediated activation of rat CYP3A1 gene and its modes of modulation by apolipoprotein AI regulatory protein I and v-ErbA-related protein 3. Arch Biochem Biophys 362:32–37

Okey AB (1990) Enzyme induction in the cytochrome P-450 system. Pharm Ther 45:241–298

Ourlin JC, Jounanaïdi Y, Maurel P, Vilarem MJ (1997) Role of liver-enriched transcription factors c/EBPα and DBP in the expression of human CYP3A4 and CYP3A7. J Hepatol 26:54–62

Padham ChRW, Paine AJ (1993) Altered expression of cytochrome P-450 mRNAs, and potentially of other transcripts encoding key functions, are triggered during the isolation of rat hepaqtocytes. Biochem J 289:621–624

Phuong-Van TL, Adesnik A, Ganguly S, Shaw PM (1996) Tanscriptional regulation of the CYP2B1 and CYP2B2 genes by C/EBP-related proteins. Biochem Pharmacol 51:345–356

Reid LM, Narita M, Fujita M, Murray Z, Liverpool C, Rosenberg L (1986) Matrix and hormonal regulation of differentiation in liver cultures. In: Guillouzo A, Guguen-Guillouzo Ch (eds) Isolated and cultured hepatocytes. Les Editions INSERM, John Libbey Eurotext, Paris, pp 225–258

Roberts BJ, Song BJ, Soh Y, Park SS, Shoaf SE (1995) Ethanol induces CYP2E1 by protein stabilization. J Biol Chem 50:29632–29635

Rogiers V, Vandenberghe Y, Callaerts A, Verleye G, Cornet M, Mertens K, Sonck W, Vercruysse A (1990) Phase I and phase II xenobiotic biotransformation in cultures and co-cultures of adult rat hepatocytes. Biochem Pharmacol 40:1701–1706

Rojkind M (1982) The extracellular matrix. In: Arias I, Popper H, Chacher D, Shafritz DA (eds) The liver biology and pathobiology. Raven Press, New York, pp 537–548

Rojkind M, Gatmaitan Z, MacKensen S, Giambrone MA, Ponce P, Reid LM (1980) Connective tisue biomatrix: its isolation and utilization for long-term cultures of normal rat hepatocytes. J Cell Biol 87:255–263

Rollier A, DiPersio CM, Cereghini S, Stevens K, Tronche P, Zaret A, Weis MC (1993) Regulation of albumin gene expression in hepatoma cells of fetal phenotype: dominant inhibition of HNF1 function and role of ubiquitous transcription factors. Mol Biol Cell 4:59–69

Runge D, Runge DM, Drenning SD, Bowen WC, Grandis JR, Michalopoulos GK (1998) Growth and differentiation of rat hepatocytes: changes in transcription factors HNF-3, HNF-4, STAT-3, and STAT-5. Biochem Biophys Res Commun 250:762–768

Santhosh A, Mathew S, Sudhakaran PR (1998) Modulation of biochemical activity of hepatocytes in culture by matrix substratum. Indian J Biochem Biophys 35:200–207

Sato Y, Ochiya T, Yasuda Y, Matsubara K (1994) A new three-dimensional culture system for hepatocytes using reticulated polyurethane. Hepatology 19:1023–1028

Schmoelzl S, Benn SJ, Laithwaite JE, Greenwood SJ, Marshall WS, Munday NA, FitzGerald DJ, LaMarre J (1998) Expression of hepatocyte low-density lipoprotein receptor-related protein is post-transcriptionally regulated by extracellular matrix. Lab Invest 78:1405–1413

Silva JM, Morin E, Day SH, Kennedy BP, Payette P, Rushmore T, Yergey JA, Nicoll-Griffith DA (1998) Refinement of an in vitro cell model for cytochrome P450 induction. Drug Metab Dispos 26:490–496

Smas CM, Sul HS (1997) Molecular mechanisms of adipocyte differentiation and inhibitory action of pref-1. Crit Rev Eukaryot Gene Expr 7:281–298

Späth GF, Weiss MC (1997) Hepatocyte nuclear factor 4 expression overcomes repression of the hepatic phenotye in dedifferentiated hepatoma cells. Mol Cell Biol 17:1931–1932

Späth GF, Weiss MC (1998) Hepatocyte nuclear factor 4 provokes expression of epithelial marker genes, acting as a morphogen in dedifferentiated hepatoma cells. J Cell Dif 140:935–946

Suolinna EM (1982) Xenobiotic metabolism and toxicity in primary monolayer cultures of hepatocytes. Med Biol 60:237–254

Tateno C, Yoshizato K (1998) Growth and differentiation of adult rat hepatocytes regulated by the interaction between parenchymal and non-parenchymal liver cells. J-Gastroenterol Hepatol 13 Suppl:S83–92

te Velde AA, Ladiges N, Flendrig LM, Chamuleau RA (1995) Functional activity of isiolated pig hepatocytes attached to different extracellular matrix substrates. Implications for application of pig hepatocytes in a bioartificial liver. J Hepatol 23:184–192

Yamada KM (1982) Biochemistry of fibronectin. The glycoconjugates, vol III. Academic Press, London, pp 331–361
Yokomori N, Nishio K, Aida K, Negishi M (1997) Transcriptional regulation by HNF-4 of the steroid 15alpha-hydroxylase P450 (Cyp2a-4) gene in mouse liver. J Steroid Biochem Mol Biol 62:307–314

Collagen Type I: A Substrate and a Signal for Invasion

Leen Van Hoorde[1,3], Elisabeth Van Aken[2,3], and Marc Mareel[1]

1
Introduction

The phenomenon termed invasion is observed in a large variety of cells and organisms from viruses to vertebrate cells (Leroy et al. 1997). Invasion implicates entry of foreign cells into a host by attachment to the extracellular matrix (ECM) components, proteolysis of these components, and subsequently migration through the ECM defects (Bruyneel and Mareel 1981; Liotta et al. 1983). Noncancer cells invade the ECM in physiological and pathological situations, for example, during embryogenesis, during wound healing, and when cells are recruited into inflammatory regions.

Cancer cells invade the ECM during the metastatic process. Cancer cells that form a noninvasive carcinoma in situ acquire an invasive phenotype upon contact with host cells and ECM components. Invasiveness of cancer cells is associated with penetration of the basement membrane and migration into the interstitial stroma, containing stromal cells and matrix molecules. Metastatic tumor cells invade lymphatic and blood vessels, a process called intravasation and extravasation, to invade another tissue (Mareel et al. 1993; Fig. 1).

Neoangiogenesis is positively correlated with cancer cell invasion (Liotta et al. 1991). During tumor angiogenesis, capillary sprouting precedes cancer cell invasion (Bajou et al. 1998) and blockage of angiogenesis suppresses invasion (Skobe et al. 1997). During early phases of angiogenesis, resting endothelial cells are stimulated to degrade the parental vessel basement membrane, to migrate into the perivascular stroma, and to initiate a capillary sprout towards the angiogenic stimulus. Lateral proteolysis of the stroma permits expansion of the sprout and lumen formation.

Morphogenesis involves the dynamic rearrangement of cells and cell layers (Takeichi 1995), a process resembling cancer invasion (Mareel et al. 1994). Solitary invasion takes place when individual mesenchymal cells leave the ectoderm at the deep surface of the primitive streak of the gastrulating embryo

[1]Laboratory of Experimental Cancerology, Department of Radiotherapy and Nuclear Medicine, Ghent University Hospital, De Pintelaan 185, 9000 Gent, Belgium
[2]Department of Ophthalmology, Ghent University Hospital, De Pintelaan 185, 9000 Gent, Belgium
[3]Both authors contributed equally to this work

Progress in Molecular and Subcellular Biology, Vol. 25
A. Macieira-Coelho (Ed.)
© Springer-Verlag Berlin Heidelberg 2000

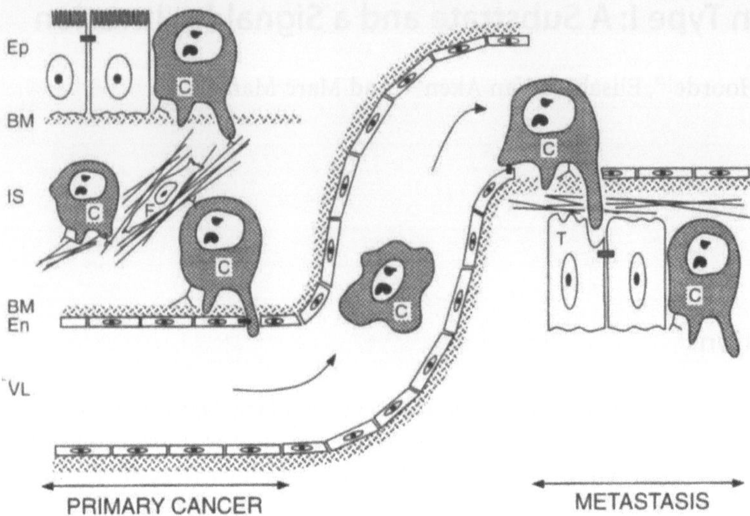

Fig. 1. Multistep invasion process of metastasis with different microecosystems of cancer invasion. *BM* Basement membrane; *C* cancer cell; *En* endothelium; *Ep* epithelium; *F* fibroblasts; *IS* interstitial stroma; *VL* vascular lumen. (After Leroy et al. 1997)

and migrate between the ectoderm and endoderm (Hatta and Takeichi 1986). Massive invasion occurs when epidermal cells bud from the mammary ridges into the surrounding mesenchyme to form the mammary gland (Sakakura 1991).

During the process of invasion in cancer, angiogenesis or morphogenesis cells must cross the barrier of the ECM. Interestingly, such a barrier, consisting of basement membrane and interstitial stroma, is produced by the potentially invading cells, by the host cells, or by both independently or under each others' influence. ECM is not only a barrier, as it serves also as a passive substrate for migrating cells and exerts signaling functions either by itself or as a molecular bulletin board retaining soluble factors (Nathan and Sporn 1991; Mareel et al. 1993). The interstitial matrix, associated with stromal fibroblasts and myofibroblasts, contains collagen type I and III, fibronectin, proteoglycans, and glycoproteins. Collagen type IV, laminin, entactin, and certain proteoglycans are uniquely localized in the basement membrane that separates organ parenchymal cells from the interstitial stroma (Liotta et al. 1983).

To study invasion in vitro, different experimental assays have been used (Mareel et al. 1991). These invasion models consist of the confrontation of potential invaders with the host to be invaded. The host is chosen to mimic the natural situation. When the host is a living organ or tissue, such as in the embryonic chick heart invasion assay, its active participation is easily taken into consideration. For "dead" matrices such as collagens or laminin this consideration was less obvious. The concept developed by Bissell et al. (1982) has

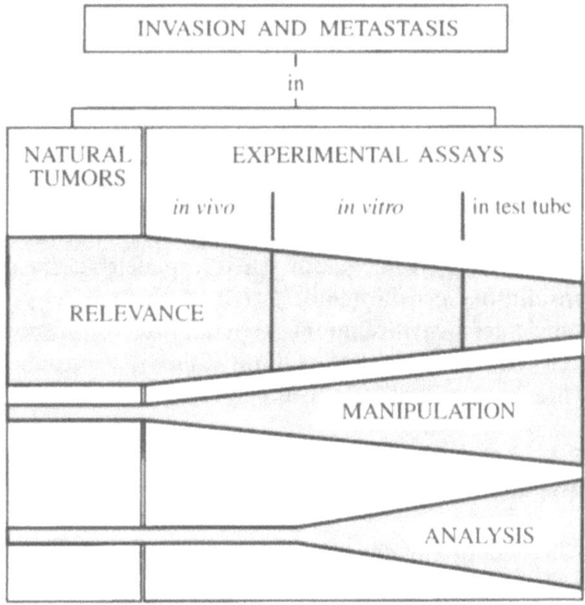

Fig. 2. The number of elements (ECM components, host cells, etc) used in in vitro invasion assays determines the relevance of observation as well as the difficulty in manipulation and analysis of the experiments. (After Mareel et al. 1991)

dramatically changed this attitude: "dead" hosts do have a profound effect on the activities of cells, including invasion. The relevance of invasion assays for the natural situation has been a matter of debate, because similarities between various assays were considered rather than differences. It is now largely accepted that invasion occurs within the frame of a microecosystem, all elements of which may influence invasion (Mareel et al. 1993) and metastasis (Fidler 1990). In vivo the number of elements present in the microecosystem of invasion is high, comprising the various normal or altered components of the ECM as well as host cells such as fibroblasts, myofibroblasts, endothelial cells, immunocytes, and inflammatory cells. Model systems, in vivo or in vitro, try to separate these elements and to bring them back together one by one, in order to understand their interaction with one another in their participation at invasion (Fig. 2).

It is the aim of the present chapter to discuss the participation of the ECM, in particular collagen type I, at invasion as a barrier, a substrate, and a signal.

2
Invasion into Matrices

The ECM is built by specific molecules. Collagen type IV, collagen type V, laminin, entactin, and heparan sulfate proteoglycans build the basement membrane. Collagen types I and III, elastin, proteoglycans, glycoproteins, and vitronectin build the interstitial stroma. Fibronectin is found in both ECMs (Liotta et al. 1983). To study invasion, gels of single components are used as well as composite gels or organ fragments. Extensively used is Matrigel, a soluble reconstituted basement membrane, extracted from the Engelbreth-Holm-Swarm mouse sarcoma tumor (Kleinman et al. 1986). It consists not only of the above mentioned components but also contains growth factors such as transforming growth factor-beta (TGF-β), fibroblast growth factor-2 (FGF-2), epidermal growth factor (EGF), platelet-derived growth factor (PDGF), insulin-like growth factor-1 (IGF-1), and nerve growth factor (NGF). Embryonic chick heart fragments are composed of myocytes, fibroblasts, endothelial cells, and components of both interstitial stroma and basement membrane (Bracke et al. 1987; De Bruyne et al. 1988).

2.1
The Structure of Collagen Type I

The structure of collagen type I is of particular interest for the present chapter (Fig. 3).

A collagen type I molecule consists of two α chains: the α1 type I or α1(I) chain, which is encoded by the *COL1A1* gene located at 17q21–22, and the α2 type I or α2(I) chain, which is encoded by the *COL1A2* gene located at 7q21–22. Each type I molecule is composed of two α1(I) chains and one α2(I) chain, the complete molecule being abbreviated $[\alpha1(I)]_2\alpha2(I)$ (Linsenmayer 1981; Murray and Keeley 1993). For experimental use, collagen type I is extracted from skin and tendons of rats, guinea pigs, and calves. The extraction occurs with neutral salt buffers, and with diluted acidic solutions or pepsin. Extracts yield milligram quantities of type I collagen, mainly in the form of monomers, i.e., single α1(I) and α2(I) chains; they contain variable amounts of cross-linked components. Since the cross-links containing terminal extensions of the collagen molecules are not in the triple helical conformation, they are susceptible to proteolytic degradation by pepsin. Vitrogen is pepsin-solubilized collagen, containing more than 90% of the type I collagen in addition to collagen types III and V. The yield of collagen extraction is enhanced when the animals are rendered lathyritic by a dietary supplement of 2-aminopropionitrile, an inhibitor of the lysine oxidase enzyme. The collagen derived from these animals has a much decreased cross-linking capacity, and contains mainly monomers. Collagen derived from nonlathyritic animals is digested by pepsin (Birkedal-Hansen 1987). Collagen fibril formation is basically a self-assembly process in vivo and in vitro. In vivo collagen type I contains unipolar as well as bipolar D-periodic fibrils. In vitro, however, this results in unipolar D-periodic fibrils containing collagen molecules which point exclusively to one direction (Kadler et al. 1996). When preparing a 3-D collagen type I gel through a cold neutral step, nonbanded filaments occur during the early phase of self-assembly. Neutralizing and warming of the collagen solution to temperatures

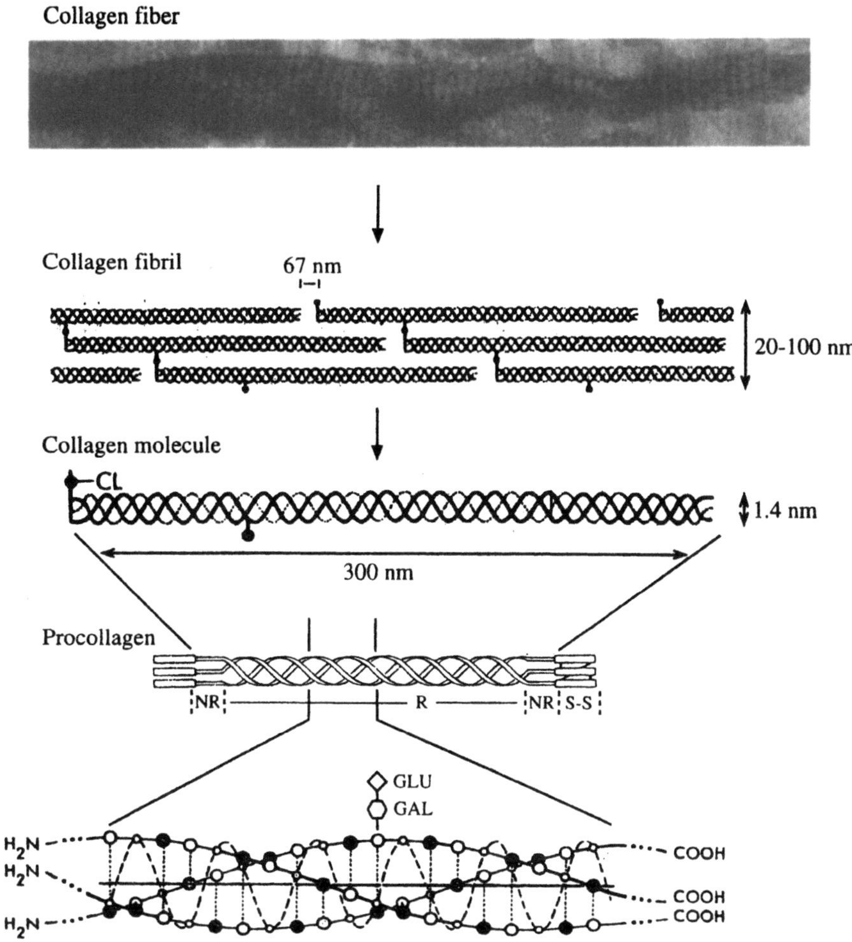

Fig. 3. Structure of collagen type I. A collagen molecule consists of three polypeptide units called α chains. Each α chain consists of a repeating (*R*) amino acid sequence (Gly-X-Y) part in the middle and a nonrepeating (*NR*) part at the amino and carboxy terminal end of the α chain, and is twisted into a left-handed helix. Glycine is presented by a *small white circle*, X by a *large shaded circle* and Y by a *large white circle*. Three of these α chains align when disulfide bonds are formed at the C-terminal ends (*S-S*), and wound into a right-handed super-helix, forming a rod-like collagen molecule 1.4 nm in diameter and about 300 nm long. The triple helix is stabilized by interchain hydrogen bonds (*straight dotted lines*). About 100 of the X positions are proline and about 100 of the Y positions are hydroxyproline. Lysines in the Y position may be posttranslationally modified to hydroxylysine. Some of these hydroxylysines may be further modified by the addition of galactose (GAL) or galactosyl-glucose (GLU). The collagen molecules self-assembly into a "quarter-staggered" alignment such that each collagen molecule is displaced longitudinally from its neighboring collagen molecule by slightly less than one-quarter of its length (67 nm). In this way, the molecules form striated fibrils of 20–100 nm in diameter, which may aggregate to form larger collagen fibers. The lysine oxidase enzyme stabilizes the collagen fibers by the formation of covalent cross-links (*CL*) between lysines and hydroxylysines. These cross-links are situated both within and between the collagen molecules and greatly decrease the solubility of the collagen molecules. (After Buddecke 1984)

between 20 and 34 °C produces a gel of D-periodic fibrils over the course of several hours. The rate of assembly into fibrils is proportional to the amount of fibrillar material formed (Kadler et al. 1996). A 3-D collagen gel shows a dense meshwork of fibers varying between 90 and 400 nm in diameter, with no particular areas of alignment or aggregation. The recognition sites in collagen type I for $\alpha_1\beta_1$ integrin are found in the CB(cyanogen bromide)3 and the CB8 fragment of the $\alpha 1$(I) collagen chain, while the recognition site of $\alpha_2\beta_1$ integrin is found only in the CB3 fragment (Gullberg et al. 1992). $\alpha_1\beta_1$ and $\alpha_2\beta_1$ integrin bind to native collagen type I in an RGD (Arg-Gly-Asp)-independent manner. Heat-denatured collagen type I, called gelatin, is bound via an RGD-sequence to $\alpha_v\beta_3$ integrin. The RGD sequences in the $\alpha 1$(I) collagen chain are located at the following amino acid positions: 564–562, 745–747, 1093–1095 and in the $\alpha 2$(I) collagen chain at the following amino acid positions: 413–411, 476–474, 777–779, 1005–1007, 1067–1065. Gelatin is obtained by heating native collagen type I until the triple helices of the collagen molecules unfold. The denaturing temperature for collagen is around 40 °C, is higher for fibrillar collagen organized in a gel than for collagen in solution, and depends on the species.

2.2
Technical Aspects: Different Collagen Type I Invasion Assays

During the process of invasion, cancer cells traverse two types of ECM: basement membranes and interstitial stroma. During metastasis cancer cells penetrate the ECM at each invasion step (Fig. 1). Different in vitro assays have been designed to mimic particular invasion steps in the process of metastasis. The Matrigel assay mimics invasion through basement membranes, the collagen type I assay invasion through interstitial stroma. Collagen invasion assays may be categorized as to the number of elements that are brought together to initiate the microecosystem (Fig. 4).

In some of these assays, cells are at the medium/collagen interface so that only their lower side feels the collagen. This type of assay is the most common and has been used to study invasion of cancer and noncancer cells (Schor 1980) as well as angiogenesis (Montesano and Orci 1985). In other assays, cells are incorporated in the gel, so that all sides of the cell feel collagen. Mixing of single cell suspensions with the collagen before gelation was used for the study of morphogenesis (Montesano et al. 1991), angiogenesis (Ilan et al. 1998), invasion (Klein et al. 1991; Oft et al. 1996), and fibroblast-induced collagen contraction, the latter with floating gels (Grinnell 1994). To assess the invasion of astrocytoma cells, multicellular spheroids were mixed with the collagen before gelation (Tamaki et al. 1997). In the collagen type I sandwich assay for angiogenesis (Jackson et al. 1992) and morphogenesis (Delannoy-Courdent et al. 1998), cells are allowed to establish a monolayer on top of the gel before addition of a second layer of collagen.

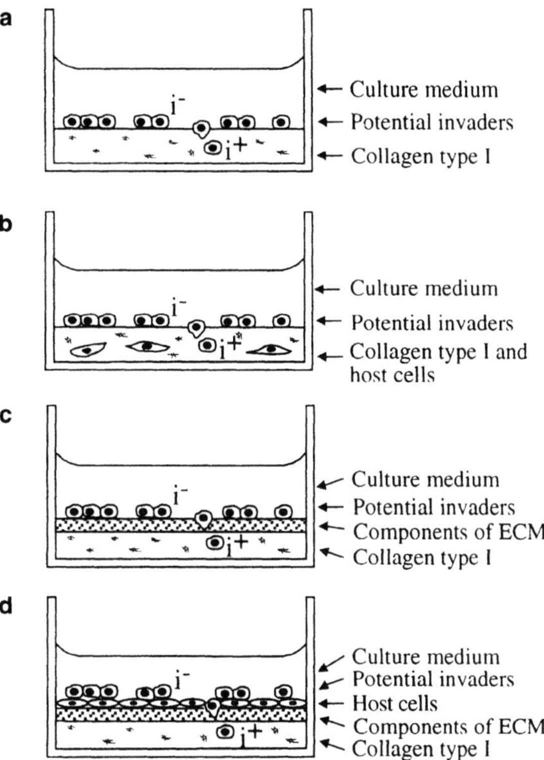

Fig. 4a–d. Schematic representation of in vitro invasion assays with the potential invaders seeded on top of the collagen type I gel. *Top (a) to bottom*: Increasing number of elements participating at the microecosystem of invasion; components of ECM are other than collagen type I. Enrichment of the gel with either host cells present in the interstitial stroma (**b**), or with other components of the ECM (**c**) or with both (**d**) was performed to analyze fibroblast-induced invasion of squamous carcinoma cells (Iwazawa et al. 1996; Yamada et al. 1999) as well as interactions between osteoblast-like cells and cancer cells derived from bone-seeking solid tumors. i^- noninvaded cells; i^+ invaded cells. (Koutsilieris et al. 1994; Sourla et al. 1996)

Addition of components of the basement membrane, such as collagen type IV, laminin, or Matrigel, or of dermal explants, was done to mimic within a single system different invasion steps of the metastatic process (Schor et al. 1985a; Muir 1995; Rosenthal et al. 1998). Nakayama et al. (1996) enriched the collagen type I gel with endothelial cells as well as collagen type IV, to investigate extravasation of cells as well as their invasion of the perivascular interstitial stroma. Obviously, the more complex the microecosystem, the more closely it mimics the natural situation, and the more relevant are the conclusions. Simplified collagen assays, however, have the advantage of a greater possibility of manipulation and greater accessibility to analysis (Fig. 2).

Invasion of cells seeded on top of collagen type I is scored under the microscope quantitatively or qualitatively on living cultures, as originally described by Schor (1980). The number of cells on top of the gel and below the gel surface is counted at approximately 10–20 regions of the gel surface, by random walk. Vakaet et al. (1991) developed a computer-assisted method to count the number of invasive cells as well as the depth of invasion. Cells are considered to be invasive when they are found deeper than $10\,\mu m$ beneath the surface of the gel. The total thickness of the collagen gel, about $250\,\mu m$, is scanned from top to bottom in steps of $12.5\,\mu m$. The invasion index is computed as the percentage of invaded cells over the total number of cells (Bracke et al. 1999). Next to the invasion index the programme permits the depth profile to be plotted, i.e., the number of cells per level scanned. Whereas the invasion index describes the proportion of cells that have invaded, the depth profile reflects the speed and the direction of invasion. An invasion index was determined by Schor (1980) removing the noninvaded cells by trypsinization and recovering the invaded cells by collagenase treatment. Cells were counted by an electronic particle counter and the invasion index was computed as the number of trypsinized cells $\times100$ over the number of trypsinized plus collagenase-extracted cells. Invasion of cells has been assessed in living cultures also by measuring the depth of the leading front of the cell population in the collagen gel (Docherty et al. 1989). To evaluate invasion of endothelial cells, the number of tubes is counted and the length and width of the tubes is measured in random microscopic fields (Gamble et al. 1993). Time-lapse videomicroscopy permits not only quantitation of invasion, but provides also an idea about the mechanics (Friedl et al. 1997). Collagen type I was used as a substrate for quantitation of invasion in coated Boyden chambers. Cells are seeded upon the collagen-coated filter insert. After incubation, cells and collagen are removed from the upper side of the filter and invaded cells on the lower side are counted after fixation and nuclear staining (Lochter et al. 1997). To distinguish between migration and invasion, cells were seeded on uncoated filters as compared with collagen-coated filters, both with pores of $12\,\mu m$ in diameter (Muir 1995). Here, the index of invasion is calculated as the percentage of invaded cells over the percentage of migrated cells.

Transverse histological sections are used for qualitative and quantitative analysis (Fisher et al. 1994; Iwazawa et al. 1996; Ilan et al. 1998). Scanning electron micrographs of cells exquisitely show the interaction of invading cells with the collagen fibers (Schor 1980). Criteria used by us and by others for morphogenesis and invasion from colonies inside collagen on living or on fixed cultures are illustrated in Fig. 5.

A more refined analysis of the leading edges of the branches was performed by Oft et al. (1996) and Delannoy-Courdent et al. (1998). Spiky ends are characteristic of invasion, bud-like ends of morphogenesis. Uyttendaele et al. (1998) measure the length of the branches and count the branching points in each colony in histological sections of the collagen gel, considering elongation and branching as two separate events during tubular morphogenesis. Values

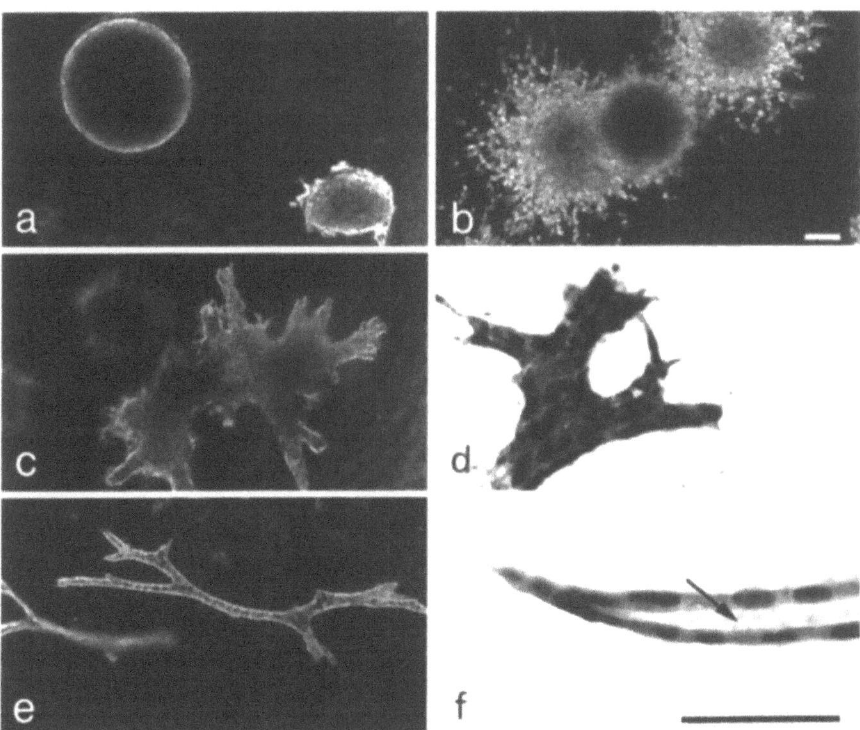

Fig. 5a–f. Morphotype of colonies grown in a three-dimensional culture in collagen type I. Morphotypes are determined on living cultures by phase contrast microscopy (**a, b, c** and **e**) or on paraffin sections stained with hematoxylin and eosin (**d** and **f**). Colonies are scored as noninvasive if they are compact with extensions less than one-tenth of the diameter of the colony (**a**). Colonies are scored as invasive if they are dispersed (**b**) or if they have extensions larger than one-tenth of the diameter of the colony (**c–f**). On histological sections the distinction is made between tubular branching, when the extensions have a lumen (indicated with an *arrow* in **f**), and solid branching, when no lumen is present (**d**). Bar 100 μm

are expressed either as mean cord length and number of branch points per photographic field or as mean cord length and number of branch points per individual colony (Soriano et al. 1996).

Several mimics of ECM remodeling during wound contraction have been developed with fibroblasts cultured in collagen type I matrices: floating matrix contraction, anchored matrix contraction, and stress relaxation. Contraction of a floating or an anchored collagen matrix results, respectively, in a mechanically relaxed or stressed tissue. Moreover, the phenotype of fibroblasts, their proliferation and response to growth factors is altered depending on anchorage of the gel (Grinnell 1994). Contraction occurs as a consequence of the motile activity of cells trying to migrate through the matrix and is assessed by measuring the diameter or surface of the gel. The index of gel contraction equals the diameter or surface of the initial gel over the diameter or surface of the contracted gel (Langholz et al. 1995; Deryugina et al. 1998).

2.3
Substrate-Dependent Differences in Invasion

Matrigel, a mimic for the basement membrane, has been compared with collagen type I, a mimic for the interstitial stroma. In these experiments, differences in collagenase expression by the cells may explain their different invasive behavior in both substrates. TCL1 and SCg6 cells derived from a mouse metastatic mammary adenocarcinoma invaded Matrigel as well as collagen type I. Invasion of Matrigel was inhibited by more than 80% when antisense oligodeoxynucleotides against stromelysin 1 were added, whereas invasion of collagen was inhibited by only 50%. This is in line with the fact that stromelysin 1 is most effective in the degradation of basement membrane constituents but only weakly attacks stromal collagens (Lochter et al. 1997). Tumor necrosis factor-α (TNF-α) stimulated invasion of keratinocytes in collagen type I and not in Matrigel (Schirren et al. 1990). Previously, it was shown that in human skin fibroblasts TNF-α induced expression of mRNA of a collagenase that does not cleave collagen type IV. Spinocellular carcinoma cells of the hypopharynx that express type I and not type IV collagenase, invaded collagen type I and not Matrigel (Wach et al. 1996). Two human melanoma cell lines derived from a primary tumor and two derived from metastases showed invasion of collagen type I, whereas four other cell lines did not invade (Wach et al. 1996). Despite the fact that all these cell lines expressed collagenase type IV in the same level, they were not all invasive in Matrigel, confirming that other factors are implicated in invasion through different substrates. The finding that some cell lines derived from metastases do not invade Matrigel can be interpreted as loss of invasiveness at the site of metastasis or in vivo assistance of invasion by factors lacking in vitro.

After 4β phorbol 12-myristate 13-acetate (PMA) treatment, a highly metastatic variant of a human colorectal carcinoma cell line was three times more invasive in Matrigel than the parental cell line (Komada et al. 1993). In the collagen type I invasion assay the parental cell line scored higher than the highly metastatic one. Two metastatic variants of a mouse lymphoma cell line invaded Matrigel but only the most metastatic one was invasive in collagen (Erkell and Schirrmacher 1988). The latter authors concluded that the rapid dissemination in vivo is due to the cancer cells' capacity to invade the interstitial stroma as well as basement membranes. It is the opinion of Hajitou et al. (1998) that enhanced invasion in collagen type I is more related to cell motility than to proteolytic activity. Murine myoepithelial cells EF43 transfected with FGF-3 invaded Matrigel while cells transfected with FGF-4 invaded collagen type I. FGF-3 expression was correlated with enhanced secretion of pro-MMP-9 and pro-MMP-2 as well as of plasminogen activators. Matrigel is able to activate secreted pro-MMPs. In contrast, FGF-4-transfected cells secreted neither pro-MMPs nor plasminogen activators, and had the largest plasminogen-activator-inhibitor activity. These results show that FGF-3 and FGF-4 induce invasion of EF43 cells via transcriptional activation of different genes, explaining the different invasive behavior in Matrigel as compared to collagen.

Positive correlation between invasion in collagen type I and in Matrigel have been reported also. Formation of arborizing structures from human gallbladder carcinoma colonies in collagen under stimulation with scatter factor/hepatocyte growth factor (SF/HGF) correlated with enhanced invasion in Matrigel (Li et al. 1998). Early-stage malignant human breast carcinoma cells were more invasive than late-stage ones in both collagen type I and Matrigel (Le Marer and Bruyneel 1996). Despite the fact that invasion in Matrigel and collagen type I is correlated, the manner of invasion may be matrix type-dependent, as observed with human lung cancer cells (Strohmaier et al. 1996). After seeding on Matrigel, a network of cells appeared on the surface of the gel and cords of cells but no single cells invaded the Matrigel. On collagen, cells remained individual and invaded the gel as single cells. Similarly, colorectal carcinoma cells SW1222 and HRA19 organized as gland-like structures in both collagen and Matrigel (Del Buono et al. 1991). Addition of an RGD-containing peptide prevented the formation of gland-like structures in collagen and not in Matrigel, indicating that the formation of gland-like structures is regulated by different mechanisms in different substrates.

Stadler and Dziadek (1996) described that choroid plexus epithelial cells invade Matrigel but not mixtures of Matrigel and collagen type I. However, in such mixtures invasion is induced by adding laminin, nidogen, or a mixture of both. The authors infer from their experiment that failure of invasion into a matrix is, at least partly, due to its molecular composition and not to physical constraints of this matrix. The influence of basement membrane components on invasion has been analyzed by a number of authors, adding to the collagen gel single components such as laminin, collagen type IV, or fibronectin. Thus, addition of laminin, but not fibronectin, to a collagen type I gel led to tube formation by mouse brain or human umbilical vein endothelial cells (Kubota et al. 1988; Kanda et al. 1999). It was suggested that $\alpha_5\beta_1$ integrin ligated to fibronectin inhibits invasion of the endothelial cells, while $\alpha_3\beta_1$ integrin ligated to laminin stimulates invasion. Overlaying collagen type I with endothelial cell ECM or fibroblast ECM inhibited the invasion of melanoma cells (Schor et al. 1985a). The invasion of chicken retinal pigment epithelial (RPE) cells into collagen type I was prevented by layering laminin and collagen type IV on the gel (Docherty et al. 1987) in line with the situation in vivo, where RPE cells reside on a basement membrane called Bruch's membrane. Sander et al. (1998) compared invasion of dog kidney MDCK-f3 cells transfected with Tiam-1 (T-cell invasion and metastasis-1) in collagen type I, laminin, and fibronectin. At low seeding density the transfected cells were less invasive in fibronectin and laminin than the parental cells. In collagen, transfectants showed an increased invasion. At high cell density, however, transfected cells showed a decreased invasion in collagen as compared to parental cells. The authors conclude that invasion in collagen is stimulated by Tiam-1 but only when cells are seeded at a low density, so that E-cadherin-mediated homotypic cell-cell adhesion is prevented. Localization of Tiam-1 is substrate-dependent: at adherens junctions in nonmotile epitheloid cells on fibronectin and laminin; in lamellae and membrane ruffles in migratory fibroblastoid cells on collagen type I.

Different integrins are implicated in branching morphogenesis in different substrates. Human mammary epithelial cells HB2 formed branching colonies under stimulation with SF/HGF in collagen type I gels as well as in fibrin gels (Alford et al. 1998). In collagen, branching was blocked with an antibody against α_2 integrins, in fibrin with an antibody against α_v integrins. This was shown also for angiogenesis (Gamble et al. 1993). Antibodies against $\alpha_v\beta_3$ integrins enhanced 4β phorbol 12-myristate 13-acetate (PMA)-induced tube formation in fibrin but not in collagen gels; antibodies against $\alpha_2\beta_1$ integrins enhanced PMA-induced tube formation in collagen gels and not in fibrin gels. HUVEC (human umbilical cord endothelial cells) formed tube-like structures in a fibrin sandwich as well as in a collagen type I sandwich (Bach et al. 1998). An antibody against VE-cadherin interrupts tube formation in both assays.

A few laboratories have used the embryonic chick heart organ culture invasion assay; here also, comparisons were made with the collagen type I assay. Malignant human breast carcinoma cells that were invasive in collagen failed to invade in the chick heart (Le Marer and Bruyneel 1996). De Ridder et al. (1994) found that nine of ten brain tumor-derived cells invaded collagen, independently of the clinical malignancy of their tumor of origin. By contrast, invasion of chick heart correlated well with malignancy in vivo. The authors suggest that in the collagen assay, cell motility is crucial, while the microenvironment of the chick heart reveals the invasive character of brain tumor cells. The invasive potential of NBT-II rat bladder carcinoma cells was demonstrated in both the collagen and chick heart assays (Tucker et al. 1990). Variations between various clones of SH27 dog mammary tumor cells, in collagen invasion but not in chick heart invasion, were correlated with E-cadherin-dependent aggregation (Spieker et al. 1995). The authors' interpretation was that factors in the microenvironment of the tumor cell determine to what degree E-cadherin is functional. The invasion suppressor role of αE-catenin, a molecule of the E-cadherin/catenin complex, is also determined by host factors. HCT-8/E11 colon cancer cells that express αE-catenin are noninvasive in chick heart and collagen, while HCT-8/E11R1 cells that lack αE-catenin are invasive in chick heart but not in collagen. Invasion of both cell lines could be induced in collagen type I by adding myofibroblasts to the collagen gel. Also in vivo both cell lines are invasive when they are injected in the colon of nude mice (our unpubl. results). This shows that host factors and modified ECM are important in determining the invasive behavior of colon cancer cells.

In summary, comparison of invasion in different substrates shows that the invasive phenotype is largely determined by host elements. Some cells invade better in collagen type I, whereas other cells invade better in other substrates. Mixing laminin in collagen type I can stimulate invasion, while putting a layer of laminin and collagen type IV on the collagen might inhibit invasion.

3
Effects of Collagen on the Invasive Behavior of Cells

3.1
Collagen Conformation

As described earlier, collagen type I occurs in different conformations: native or triple helical collagen, denatured collagen lacking triple-helical domains, and monomeric collagen, which is triple-helical but does not form fibrils (see Fig. 3). These conformational differences might influence the invasive behavior of the cells.

Sweeney et al. (1998) showed that addition of collagen type I to a monolayer of endothelial cells gave rise to formation of endothelial tubes only when native collagen but not when denatured collagen was used. Denaturing of the collagen occurs during invasion of melanoma cells by local damage of the ECM due to release of proteases (Davis 1992). The formation of denatured collagen uncovers cryptic Arg-Gly-Asp (RGD)-binding sites. The RGD-binding integrin $\alpha_v\beta_3$ of melanoma cells was found to bind strongly to denatured collagen and weakly to native collagen. The expression of $\alpha_v\beta_3$ integrin increases during the progression of melanoma cells to a more malignant and metastatic phenotype, suggesting that this integrin is implicated in the progression of melanoma. Denaturing of collagen uncovers also a cryptic RGD-containing site recognized by the $\alpha_5\beta_1$ integrin of hepatocytes (Gullberg et al. 1992). Here also, the RGD sequence involved in the binding of integrins to collagen is functional when the collagen is in the denatured form but not in the triple-helical native conformation.

Schor et al. (1996) examined the effects of native fibronectin and several of its principal functional domains on the invasion of fibroblasts into native collagen type I matrices as compared to gelatin-coated polycarbonate filters by adding native fibronectin or one of its fragments. The gelatin-binding domain of fibronectin stimulates invasion in native collagen type I-coated filters but not in gelatin-coated filters. The initial interaction of the gelatin-binding domain with its putative cell surface receptor is not dependent upon the collagen substratum, whilst the enhanced invasion is substrate-dependent. Gelatin-coated filters may be too unstable to support migration, fibroblasts needing a more rigid structure to migrate (Docherty et al. 1989). The authors claim that the use of a 3-D collagen matrix instead of a gelatin-coated filter is a more physiologically relevant assay.

In a glycosylated 3-D collagen lattice endothelial cell invasion was less pronounced than in a nonglycosylated gel (Kuzuya et al. 1998). Glycosylation is associated with structural alteration of matrix components, such as collagen-to-collagen cross-linking, which leads to an increased stiffness of matrix fibers and a reduced sensitivity to proteolytic enzymes. Following 2 days of incubation in the presence of angiogenic factors, a decreased number of invading endothelial cells was observed in glycosylated as compared to nonglycosylated collagen.

The concentration of the collagen gel may influence the degree and the way of invasion. Gels at less than 0.5 mg collagen ml^{-1} were too unstable to support fibroblast invasion, whereas gels at more than 3.5 mg ml^{-1} were too concentrated to permit it (Docherty et al. 1989). Maximal invasion of L fibroblasts occurred at collagen concentrations between 1 and 2 mg ml^{-1} (Chen and Öbrink 1991). Also for melanoma cells, collagen matrices at concentrations between 0.67 and 2.5 mg ml^{-1} are permissive for migration (Schor et al. 1982). Based on the latter observations, Friedl et al. (1997) established a collagen model for invasion of MV3 melanoma cells with a concentration high enough to provide a certain degree of physical resistance, i.e., with a pore diameter smaller than the approximate diameter of the migrating cells. Melanoma cells invading a collagen matrix at 1.67 mg ml^{-1} had to rearrange the collagen lattices, since the pore diameter of the matrix was 66.8 µm, whereas the mean diameter of the migrating cell clusters was 87 µm. Penetration of endothelial cells into highly malleable gels of 0.2 mg ml^{-1} collagen was extensive after a relatively short, 2-day, period of culture but declined in gels with higher concentrations of collagen (Vernon and Sage 1999). Sprouts that penetrated gels of 0.2 mg ml^{-1} exhibited little branching. In contrast, sprouts formed in gels of intermediate density, i.e., 0.6 mg ml^{-1}, branched and associated to form networks and arcades. Sprouts developed in rigid cells of 2 mg ml^{-1} branched extensively and formed large aggregates at the invasion front. The authors suggest that collagen gels of low concentration that permit a high degree of endothelial cell invasion would be optimal for evaluation of potential inhibitors of angiogenesis, whereas more concentrated gels would be suited to study compounds that stimulate vascular growth.

3.2
Organization of the Invader/Collagen Microecosystem

Cell density migration index was introduced because, in many experiments, the seeding density of the cells seemed to influence their invasion into collagen type I (Schor et al. 1985b). For example, fetal and cancerous fibroblasts seeded at the higher density of 2.5×10^4 cells cm^{-2} invaded more than normal adult fibroblasts, whereas at the lower density of 1×10^3 cells cm^{-2} adult fibroblasts were the more invasive cells (Schor and Schor 1987). In angiogenesis assays, a high inoculum of endothelial cells, provoking cell-cell contact, is needed for tubule formation (Kanda et al. 1999). Mammary tumor spheroids, but not single cells, invaded collagen (Zamora et al. 1980). L fibroblasts transfected with E-cadherin were not able to penetrate into collagen at high seeding density; at lower seeding density, E-cadherin-positive and E-cadherin-negative cells invaded equally well (Chen and Öbrink 1991). The most plausible explanation for this density-dependent inhibition of invasion is E-cadherin-mediated cell-cell adhesion, operating in confluent cultures. On the other hand, Meyer et al. (1995) showed that invasion of NCAM-expressing L fibroblasts in collagen was inhibited at sparse as well as at confluent cell densities. Since there was no difference in invasion depending upon cell density, the authors claimed that NCAM-mediated inhibition of invasion was not due to cell-cell interactions but rather to cell-collagen interactions mediated by NCAM.

The position of the cells with regard to the collagen, i.e., seeded upon a collagen layer or mixed in the gel may influence the behavior of the cells. Martel et al. (1997) found that invasion of MDCK(LT) cells into collagen type I was correlated with formation of tube-like structure in a 3-D collagen gel. In our own experiments we found that myofibroblast-induced invasion of HCT-8 colon cancer cells in collagen was correlated with formation of branching structures of HCT-8 cells in a 3-D collagen culture under influence of myofibroblast-conditioned medium. The same was observed for NMuMG and MMT mouse mammary epithelial cells by Delannoy-Courdent et al. (1998). Reduced invasion in collagen type I correlated with inhibition of tubular morphogenesis and reduced scattering in 3-D cultures. Several differences exist between 2-D and 3-D migration systems. The migration of cells across 2-D planar substrates is governed by a firm enough, yet transient, attachment of cells to the substrate. The migration of cells within a 3-D matrix is controlled additionally by the degree of fiber strength and density as compared to the dimensions and flexibility of the migrating cell body (Friedl et al. 1997). Microvascular endothelial cells respond differently to the presence of TGF-β when cultured on a 2-D coat of type I collagen as compared to a 3-D collagen gel (Sankar et al. 1996). Whereas cells on 2-D gels mimic the tip of an angiogenic sprout, cells in 3-D matrices mimic more distal and more differentiated parts of the sprout.

3.3
Invasion of Cells Without External Stimuli Other Than Collagen Type I

Some cells invade collagen without addition of external stimuli other than the collagen itself. To prevent contamination with growth factors or other ECM components, the collagen type I used in the different assays is purified. Upon purity check by gel electrophoresis, Schor et al. (1982) demonstrated that some batches of collagen contain proteoglycans as their major contaminant.

The invasion of cancer cells is often inversely related to the expression of E-cadherin, a well-documented invasion-suppressor molecule (Vleminckx et al. 1991). For example, Frixen et al. (1991) showed for a panel of 23 human cancer cell lines derived from bladder, colon, breast, lung, and pancreas that E-cadherin-positive cell lines were noninvasive in collagen, whereas the E-cadherin-negative cell lines did invade. This was also observed by other investigators for different kinds of carcinomas (Kinsella et al. 1994; Galzie et al.

1997), and for noncancerous endometriotic cells that also in vivo invade the peritoneum, gut, or lung (Gaetje et al. 1995, 1997).

Other factors repeatedly mentioned as mediating invasion in collagen type I are matrix metalloproteinases (MMPs). Invasion of breast carcinoma cells, osteoclasts and astrocytoma cells were prevented by MMP inhibitors, indicating that proteolytic cleavage of the collagen fibers is needed for invasion (Tamaki et al. 1997; Kelly et al. 1998; Sato et al. 1998). Enzyme specificity appeared from comparison between normal Schwann cells, producing MMP-2 and MMP-3, and astrocytoma cells, producing MMP-1 and MMP-9 (Muir 1995). The latter did invade, but the former failed to do so, in line with the fact that MMP-1 cleaves collagen type I fibrils. The manner in which cells invade from C6 astrocytoma spheroids suggests that proteases are involved (Tamaki et al. 1997). These cells invade the gel in a radial and unidirectional way and not in the random way provided by spaces available between the collagen type I fibers. Melanoma cells invade in collagen without hydrolysis of the collagen fibers (Schor et al. 1985a; Kono et al. 1990). In 3-D collagen lattices, such cells developed a slow adhesive type of migration consisting of one or two leading pseudopodia, attachment to individual collagen fibers, and contraction of the cell body (Friedl et al. 1997). Wordinger et al. (1991) showed that the morphology of collagen type I was altered in the immediate vicinity of trophoblast projections, indicating an increased proteolytic activity. Although the interstitial collagen model lacks components found in situ, the hydrated collagen gel does mimic the organizational framework of the connective tissue stroma in vivo. The in vitro findings of Wordinger et al. (1991) were validated by the experiments of Katz (1995), who showed in vivo that trophoblasts from mice that were fasted for 24h are capable of breaking down extracellular collagen. Whilst invasion of trophoblasts in the underlying stroma is a normal physiological event, invasion of RPE cells is not, except for pathological conditions such as retinal detachment. Nevertheless, Docherty et al. (1987) found that chick embryo RPE cells invaded collagen type I. Invasion occurred when RPE cultures were postconfluent and multilayered.

Fetal fibroblasts invade in collagen (Docherty et al. 1989; Kim et al. 1998). A migration-stimulating factor is produced by fetal fibroblasts and fibroblasts from cancer patients, but not by adult fibroblasts, correlating with the migration of these cells into collagen (Schor and Schor 1987).

Despite the fact that some types of cells can invade as such in collagen type I, many of them need external stimuli such as growth factors or proteases recruited from helper cells.

3.4
Invasion, Angiogenesis, and Morphogenesis Assisted by Helper Cells

In vivo the stroma contains, next to ECM such as collagen type I, also host cells such as fibroblasts and myofibroblasts. The production of collagen type I is the result of a cross-talk between epithelial cells and stromal fibroblasts. Some cells that are unable to invade pure collagen, do invade once fibroblasts or myofibroblasts are incorporated in the gel (Table 1).

Sometimes conditioned medium of the cells was equally efficient, indicating secretory stimulators of invasion. One of these stimulators that is frequently implicated in host cell-induced cancer invasion and in morphogenesis is scatter factor/hepatocyte growth factor (SF/HGF) (Shimura et al. 1995; Soriano et al. 1995; Iwazawa et al. 1996; Bae-Jump et al. 1999). Factors implicated in endothelial cell tubulogenesis in presence of cancer cells are fibroblast growth factor (FGF)-2 and TGF-α (Ono et al. 1992; Abe et al. 1993).

Developmental regulation of the fibroblast-epithelial interactions was evident from human fetal, but not adult, lung fibroblasts inducing morphogenesis of normal human tracheobronchial epithelial cells in collagen (Infeld et al. 1993). Host cell type specificity was also observed with colon cancer myofibroblasts, but not normal colon fibroblasts, inducing invasion of colon cancer cells HCT-8 (our unpubl. results).

Invasion is influenced by interactions of cancer cells not only with stromal cells but also with local environmental factors (Chen and Tseng 1995; Yamada et al. 1999). Laryngeal carcinoma cells and normal ocular surface epithelial cells invaded more when the collagen gel was exposed to air than to liquid culture medium. Seeding of laryngeal carcinoma cells at the collagen-air interphase more closely resembles the laryngeal environment (Yamada et al. 1999). In the model developed by Koutsilieris et al. (1994), osteoblast-like cells are mixed homogeneously with the collagen gel and cancer cells are injected focally into the gel. These models permitted the analysis of cancer cell invasion and local host reaction: prostate cancer cells PC-3 invade, attract osteoblasts, and cause increased density of the collagen; breast cancer cells MCF-7 and ZR-75 invaded, decreased the collagen density, and did not attract osteoblasts; endometrial cancer cells KLE did not change the gel (Sourla et al. 1996). These observations are suggestive of a cancer cell-specific interaction with osteoblasts in this collagen type I invasion system, as they mimic the cells' propensity to produce osteolytic metastases, as for the breast cancer cells, or osteoblastic metastases, as for the prostate cancer cells.

3.5
Invasion Stimulated by Extrinsic Factors

Many types of cancer cells fail to invade collagen type I unless stimulated by extrinsic factors (Table 2). Such experiments are useful also for the analysis of factors that are implicated in angiogenesis and in morphogenesis.

Factors shown to be angiogenic in collagen type I assays, as well as in vivo, include EGF (Ono et al. 1992), FGF-2 (Montesano et al. 1986; Gajdusek et al. 1993; Goto et al. 1993; Kanda et al. 1999), FGF-4 (Deroanne et al. 1997), TGF-β (Gajdusek et al. 1993; Sankar et al. 1996), and VEGF (Goto et al. 1993; Montesano et al. 1996; Vernon and Sage 1999). In the absence of angiogenic factors, only a few endothelial cells invaded and no cord formation was observed.

Table 1. Helper cells in collagen type I invasion, angiogenesis and morphogenesis

Invader cell	Helper cell	Factors	Reference
Invasion			
Breast ca (hu) MCF-7, ZR-75	Osteoblast-like cells (hu) MG-63		(1)
Colon ca (hu) HCT-8	Colon ca myofi (hu) hCT5.1		(2)
Endometrial ca (hu)			
KLE, HEC-1A	Endometrium stromal fi (hu)	SF/HGF	(3)
Esophagus squamous ca (hu)			
TE2	Esophagus ca fi (hu)	SF/HGF	(4)
	Esophagus submucosa fi (hu)		
	Fetal lung fi (hu) MRC-5	E-CAD	
Gallbladder ca (hu) GB-d1	Fetal lung fi (hu) HEL; MRC-5; TIG	SF/HGF	(5)
Larynx squamous ca (hu)			
HEp-2	Laryngeal fi (hu) + fat cells (ra)		(6)
Oral squamous ca (hu)			
Ca9-22; NA; HSC-2,-3,-4	Oral mucosa fi (hu)		(7)
OSC-19	Fetal fi (mu) 3T3	MMP-9	(8)
PE/CA-PJ15 & 41	Oral ca stromal fi (hu)		(9)
	Palmar fibromatosis fi (hu)		
Prostate ca (hu) PC-3	Osteoblast-like cells (hu) MG-63		(10)
Angiogenesis			
Aorta (bo) BAEC	Glioma cells (hu) IN301, IN500, U251	bFGF	(11)
Dermal microvascular (hu)			
HDMEC	mast cells (hu) HMC-1		(12)
Microvascular (bo) BMEC	Fetal fi (mu) 3T3	uPA	(13)
Omentum microvascular (hu)			
HOME	Keratinocytes (hu)	TGF-α	(14)
Pulmonary artery (bo) CPAE	Fetal fi (mu) 3T3	uPA	(13)
Umbilical vein (hu)			
HUVEC	Fetal fi (mu) 3T3	uPA	(13)
EA.hy926	Skin fi (ra)		(15)
Morphogenesis			
Corneal and conjunctival ep (r)	Corneal or conjunctival fi (r)		(16)
	Fetal fi (mu) 3T3		
Mammary ep (mu) TAC-2	Fetal fi (mu) 3T3	SF/HGF	(17)
	Fetal lung fi (hu) MRC-5	SF/HGF	
SMG ductal ep (mu) SIMS	SMG fi-like cells (mu) FS10		(18)
	Fetal lung fi (hu) MRC-5		
Tracheobroncheal adult ep (hu)	Fetal lung fi (hu) GM-5387; HFL-1		(19)

bo, bovine; ca, carcinoma; ep, epithelial cells; E-CAD, E-cadherin; FGF, fibroblast growth factor; fi, fibroblast; hu, human; MMP, matrix metalloproteinase; mu, murine; r, rabbit; ra, rat; SF/HGF, scatter factor/hepatocyte growth factor; SMG, submandibular gland; TGF, transforming growth factor; uPA, urokinase plasminogen activator.
(1) Sourla et al. 1996; (2) our unpubl. results; (3) Bae-Jump et al. 1999; (4) Iwazawa et al. 1996; (5) Shimura et al. 1995; (6) Yamada et al. 1999; (7) Matsumoto et al. 1989; (8) Kawahara et al. 1993; (9) Berndt et al. 1997; (10) Koutsilieris et al. 1994; (11) Abe et al. 1993; (12) Blair et al. 1997; (13) Montesano et al. 1993; (14) Ono et al. 1992; (15) Schönherr et al. 1999; (16) Chen and Tseng 1995; (17) Soriano et al. 1995; (18) Laoide et al. 1999; (19) Infeld et al. 1993.

Table 2. Extrinsic factors stimulating invasion into collagen type I

Factor	Invader cell	Reference
EGF	Skin squamous ca (hu) HSC-1	(1)
	Skin squamous ca (hu) UM-SCC-1	(2)
	Epidermal keratinocytes (hu)	(3)
	Keratocytes (hu)	(4)
FGF-2	Colon ca (hu) HCT-116A	(5)
	Keratocytes (hu)	(4)
SF/HGF	Esophageal squamous cell ca (hu) TE2	(6)
	Skin squamous cell ca (hu) UM-SCC-1	(2)
	Endometrial ca (hu) HEC-1A, KLE	(7)
	Bladder ca (ra) NBT-II	(8)
	Gallbladder ca (hu) GB-d1, GB-d2, GB-h3	(9)
	Gallbladder ca (hu) GB-d1	(10) (11)
	Colon ca transfected with *c-src* oncogene (hu) PCmsrc	(12)
	Kidney ep transfected with *src* oncogene (d) MDCKts.*src* Cl2	(12) (13)
PDGF	Keratocytes (hu)	(4)
TGF-α	Epidermal keratinocytes (hu)	(3)
TGF-β	Prostate ca (hu) PC3	(14)
	Pulmonary ca (hu) A549	(15)
	Kidney ep transfected with mutSmad2 (d) MDCK	(16)
	Mammary ep transfected with v-Ha-ras (mu) EpRas	(17)
TGF-βRII	Colon ca with mutTGFβRII (hu) DLD-1	(18)
MMPs	Breast ca (hu) MDA-MB-231	(19)
E-cadherin Ab	Salivary ca (mu) CSG 120/7	(20)
	Kidney ep (d) MDCK	(21)
	Breast ca (hu) T47D	(22)
	L fi transfected with E-cadherin cDNA (mu)	(23)
	Colorectal ca (hu) HCT-116A	(24)
	Esophageal ca (hu) ECD(+)-1TE2	(25)
sE-cadherin	Kidney ep transfected with *src* oncogene (d) MDCKts.*src* Cl2	(26)
HA/CS	Choroid fi (c)	(27)
TSP-1	Oral squamous ca (hu) KB	(28)
	Breast ca (hu) MDA-MB-231, SKBR-3, MCF-7	(29)

Ab, antibody; bo, bovine; c, chick; ca, carcinoma; d, dog; EGF, epidermal growth factor; ep, epithelial cells; FGF, fibroblast growth factor; fi, fibroblast; HA/CS, hyaluronic acid/chondroitin sulfate; hu, human; MMP, matrix metalloproteinase; mu, murine; PDGF, platelet derived growth factor; r, rabbit; ra, rat; sE-cadherin, soluble E-cadherin; SF/HGF, scatter factor/hepatocyte growth factor; TGF, transforming growth factor; TGF-βRII, transforming growth factor beta receptor type II; TSP-1, thrombospondine-1.
(1) Fujii et al. 1996; (2) Rosenthal et al. 1998; (3) Turksen et al. 1991; (4) Andresen et al. 1997; (5) Galzie et al. 1997; (6) Iwazawa et al. 1996; (7) Bae-Jump et al. 1999; (8) Bellusci et al. 1994; (9) Li et al. 1998; (10) Date et al. 1998; (11) Shimura et al. 1995; (12) Kotelevets et al. 1998; (13) Empereur et al. 1997; (14) Festuccia et al. 1999; (15) Mooradian et al. 1992; (16) Prunier et al. 1999; (17) Oft et al. 1996; (18) Oft et al. 1998; (19) Kelly et al. 1998; (20) Frixen et al. 1991; (21) Behrens et al. 1989; (22) Frixen and Nagamine 1993; (23) Chen and Öbrink 1991; (24) Kinsella et al. 1994; (25) Doki et al. 1993; (26) Noë et al. 1999; (27) Docherty et al. 1989; (28) Wang et al. 1995; (29) Wang et al. 1996.

The main factors implicated in the morphogenesis of mammary epithelial cells and kidney cells investigated in 3-D collagen cultures are SF/HGF (Montesano et al. 1991; Berdichevsky et al. 1994; Crepaldi et al. 1997; Alford et al. 1998; Hirai et al. 1998) and TGF-β (Soriano et al. 1996; Uyttendaele et al. 1998).

The influence of certain factors on invasion in collagen was also investigated through viral infection or transfection of constructs containing the cDNA of interest under the control of a constitutive or an inducible promoter. In such a manner, cells were induced to invade upon overexpression of metastasin, a Ca^{2+}-binding protein of the S100 family (Keirsebilck et al. 1998), or of the oncogenic nonreceptor tyrosine kinase SRC (Matsuyoshi et al. 1992; Behrens et al. 1993). E-cadherin is a paradigm of an invasion-suppressor molecule. Using the collagen assay, Frixen et al. (1991) found an inverse correlation between invasion and expression of E-cadherin in a variety of human cancer cells. The role of E-cadherin in invasion was further demonstrated by transfection of E-cadherin-negative or -positive cells with E-cadherin sense or antisense cDNA, respectively (Vleminckx et al. 1991). Invasion was also induced by addition of antibodies functionally neutralizing E-cadherin. Recently, Noë et al. (1999) showed that HAV (histidin-alanin-valin)-containing peptides homologous with the first extracellular domain of E-cadherin interfere with E-cadherin functions and induce invasion. A similar activity was exerted by the concentrated conditioned medium of cells that produce s(oluble) E-cadherin fragments through ectodomain shedding. The above-mentioned experiments with the invasion-suppressor molecule E-cadherin, as well as with various invasion-promoter molecules, underline the crucial role of the collagen type I assay in our understanding of the molecular mechanisms of cancer invasion. We would like to emphasize, however, that this assay covers only some aspects of the microecosystem of invasion in vivo.

4
A Scenario for the Molecular Cross Talk Between Collagen and Cells

MMPs, serving degradation of matrix components, and integrins, mediating attachment of cells to the ECM, have received major attention from investigators of the molecular cross-talk between cells and collagen type I.

In vivo, overexpression of integrins and MMPs is positively correlated with tumor invasiveness (Sato et al. 1994). $\alpha_2\beta_1$ and $\alpha_1\beta_1$ integrin-dependent remodeling of the ECM is involved in metastasis (Chan et al. 1991). Integrins function as transmembrane linkers between the ECM and the actin cytoskeleton. In the β_1 subgroup of integrins, at least two receptors, namely $\alpha_1\beta_1$ and $\alpha_2\beta_1$, interact with collagen type I. The MMPs are a family of zinc-dependent endoproteinases that hydrolyze proteins of the ECM. The interstitial collagenases, MMP-1, MMP-8, and MMP-13, belong to an MMP subfamily that specifically cleaves native triple helical collagens, yielding 3/4- and 1/4-

length collagen fragments as a result of the hydrolysis of a single Gly-Ile/Leu bond in each α chain of the collagen molecule (Birkedal-Hansen 1987). The collagen fragments produced by the interstitial collagenases are susceptible to further breakdown by the gelatinases MMP-2 and MMP-9 (Aimes and Quigley 1995).

In vitro, fibroblasts and tumor cells use $\alpha_2\beta_1$ and $\alpha_1\beta_1$ integrins for the reorganization of 3-D collagen lattices and for the migration into the collagen. Synthesis and transcription of β_1 integrin is selectively upregulated during the contraction process and returns to baseline levels after the contraction has subsided. Structural matrix organization in the tumor microenvironment may contribute to tumor cell invasion: melanoma cells show a tendency to migrate through matrix areas reorganized by previous passenger cells (Friedl et al. 1997). In this remodeling of the ECM in vivo and of the collagen in vitro, different events are implicated. A possible scenario of such events is presented in Fig. 6.

The hypothesis considers the functional link between integrin-mediated matrix adhesion and MMP proteolysis (Ellerbroek et al. 1999). Contact of cancer cells or fibroblasts with the ECM results in enhanced cell surface MMP activity. MMP-mediated proteolysis of the collagen-rich matrix decreases β_1 integrin aggregation, thereby disrupting signaling pathways required for induction of MMP activation. In this way, invasion of cells in the ECM is balanced (Ellerbroek et al. 1999). Of course, the schematic of Fig. 6 does not include all elements of the microecosystem of collagen invasion. SPARC (secreted protein acidic and rich in cysteine) is a matricellular protein that participates at collagen type I-induced MMP-2 activation through pro-MMP-2 (Gilles et al. 1998). Syndecans, a subfamily of the heparan-sulfate proteoglycans, act as invasion-suppressors. Indeed, cancer cells that express syndecan-1, syndecan-2, or syndecan-4 bind to collagen type I but fail to invade (Liu et al. 1998). Missing on our scheme are also chemotactic and haptotactic factors. Components of the ECM, including collagens, have been shown to promote directional haptotactic migration (Komada et al. 1993; Chintala et al. 1996; McCarthy et al. 1996). The migration of endothelial cells in response to collagen type I is blocked by antibodies to $\alpha_2\beta_1$ integrin (Leavesly et al. 1993). In contrast, Nabeshima et al. (1986) showed no haptotactic activity of native collagen to tumor cells but of collagen type I degradation products. They hypothesize that tumor cells release collagen type I degradation products by producing collagenase, and that the following cancer cells migrate via haptotaxis to these degradation products.

5
Conclusion

Invasion in vivo results from the molecular cross-talk between potential invaders, host cells, and the ECM, with collagen type I as a major component.

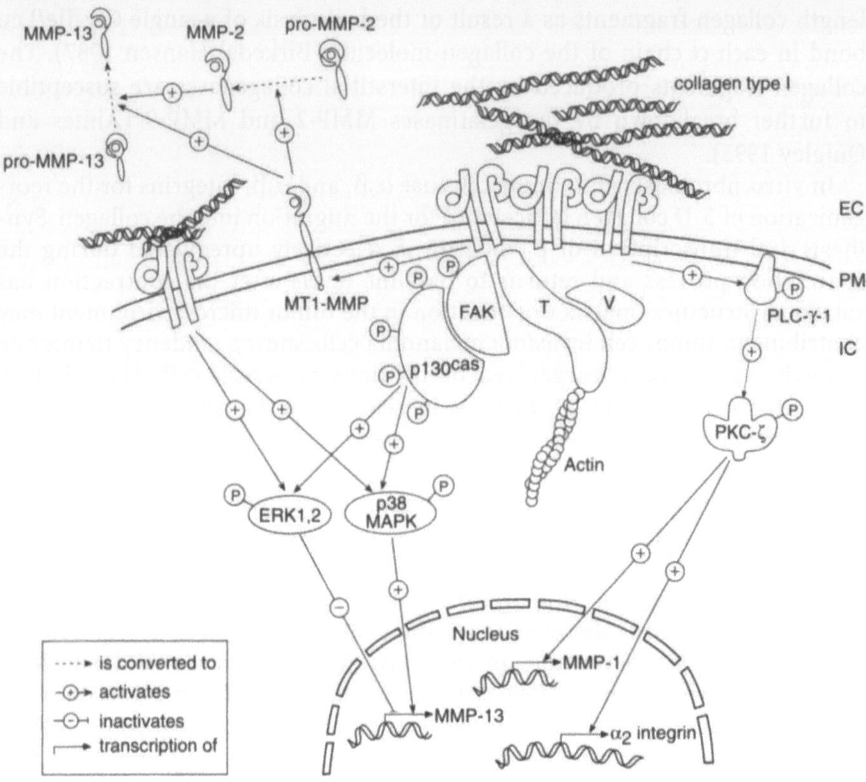

Fig. 6. Schematic representation of integrin-mediated matrix adhesion and matrixmetallo-proteinase (MMP) proteolysis. Contact with collagen type I leads to clustering and activation of integrins, followed by cascades of phosphorylation pathways and gene activation or by posttranslational modification of membrane-associated MMP (MT1-MMP). α β Integrin sub-units; *EC* extracellular; *ERK-1,2* extracellular signal-regulated kinase-1,2; *FAK* focal adhesion kinase; *IC* intracellular; *MAPK* mitogen-activated protein kinase; *P* phosphorylation; $p130^{cas}$ p130 protein crk-associated substrate; *PKC-ζ* protein kinase C-ζ; *PLCγ-1* phospholipase Cγ-1; *PM* plasma membrane; *T* talin; *V* vinculin. (After Ellerbroek et al. 1999, with data from Roeckel and Krieg 1994, Sato et al. 1994, Schlaepfer et al. 1994, Seltzer et al. 1994, Schaller et al. 1995, Knäuper et al. 1997, Langholz et al. 1997, Xu and Clarck 1997, and Ravanti et al. 1999)

Different in vitro settings containing collagen type I mimic the invasiveness in vivo. Thereby invasion varies with the geometry, the structure, and the density of the collagen. Although some cancer as well as noncancer cells invade colla-gen without external stimuli, many of them need host cells or ECM compo-nents other than collagen type I that can help to cross the collagen barrier. Collagen not only supports the invader passively, but also helps the cells to invade actively. Invasion implies a bidirectional cross-talk between the invader and collagen: both act as donor and acceptor of signals for invasion.

Acknowledgements. Financial support from the Fund for Scientific Research-Flanders (FWO, Brussels, Belgium), the Vlaams instituut voor de bevordering van het wetenschappelijk onderzoek in de industrie (IWT, Brussels, Belgium), the Sportvereniging tegen Kanker (Brussels, Belgium), the ASLK/VIVA-verzekeringen (Brussels, Belgium) and the G.O.A. from the Vlaamse Gemeenschap (Brussels, Belgium) is gratefully acknowledged.

References

Abe T, Okamura K, Ono M, Kohno K, Mori T, Hori S, Kuwano M (1993) Induction of vascular endothelial tubular morphogenesis by human glioma cells. A model system for tumor angiogenesis. J Clin Invest 92:54–61

Aimes RT, Quigley JP (1995) Matrix metalloproteinase-2 is an interstitial collagenase. Inhibitor-free enzyme catalyzes the cleavage of collagen fibrils and soluble native type I collagen generating the specific 3/4- and 1/4-length fragments. J Biol Chem 270:5872–5876

Alford D, Baeckström D, Geyp M, Pitha P, Taylor-Papadimitriou J (1998) Integrin-matrix interactions affect the form of the structures developing from human mammary epithelial cells in collagen or fibrin gels. J Cell Sci 111:521–532

Andresen JL, Ledet T, Ehlers N (1997) Keratocyte migration and peptide growth factors: the effect of PDGF, bFGF, EGF, IGF-I, aFGF and TGF-β on human keratocyte migration in a collagen gel. Curr Eye Res 16:605–613

Bach TL, Barsigian C, Chalupowicz DG, Busler D, Yaen CH, Grant DS, Martinez J (1998) VE-cadherin mediates endothelial cell capillary tube formation in fibrin and collagen gels. Exp Cell Res 238:324–334

Bae-Jump V, Segreti EM, Vandermolen D, Kauma S (1999) Hepatocyte growth factor (HGF) induces invasion of endometrial carcinoma cell lines in vitro. Gynecol Oncol 73:265–272

Bajou K, Noël A, Gerard RD, Masson V, Brunner N, Holst-Hansen C, Skobe M, Fusenig NE, Carmeliet P, Collen D, Foidart JM (1998) Absence of host plasminogen activator inhibitor 1 prevents cancer invasion and vascularization. Nat Med 4:923–928

Behrens J, Mareel MM, Van Roy FM, Birchmeier W (1989) Dissecting tumor cell invasion: epithelial cells acquire invasive properties following the loss of uvomorulin-mediated cell-cell adhesion. J Cell Biol 108:2435–2447

Behrens J, Vakaet L, Friis R, Winterhager E, Van Roy F, Mareel MM, Birchmeier W (1993) Loss of epithelial morphotype and gain of invasiveness correlates with tyrosine phosphorylation of the E-cadherin/β-catenin complex in cells transformed with a temperature-sensitive v-src gene. J Cell Biol 120:757–766

Bellusci S, Moens G, Gaudino G, Comoglio P, Nakamura T, Thiery JP, Jouanneau J (1994) Creation of an hepatocyte growth factor/scatter factor autocrine loop in carcinoma cells induces invasive properties associated with increased tumorigenicity. Oncogene 9:1091–1099

Berdichevksy F, Alford D, D'Souza B, Taylor-Papadimitriou J (1994) Branching morphogenesis of human mammary epithelial cells in collagen gels. J Cell Sci 107:3557–3568

Berndt A, Hyckel P, Könneker A, Katenkamp D, Kosmehl H (1997) Oral squamous cell carcinoma invasion is associated with a laminin-5 matrix re-organization but independent of basement membrane and hemidesmosome formation. Clues from an in vitro invasion model. Invasion Metastasis 17:251–258

Birkedal-Hansen H (1987) Catabolism and turnover of collagens: collagenases. Methods Enzymol 144:140–171

Bissell MJ, Hall HG, Parry G (1982) How does the extracellular matrix direct gene expression? J Theor Biol 99:31–68

Blair RJ, Meng H, Marchese MJ, Ren S, Schwartz LB, Tonnesen MG, Gruber BL (1997) Human mast cells stimulate vascular tube formation. Tryptase is a novel, potent angiogenic factor. J Clin Invest 99:2691–2700

Bracke ME, Castronovo V, Van Cauwenberge RML, Vakaet L Jr, Strojny P, Foidart J-M, Mareel MM (1987) The anti-invasive flavonoid (+)-catechin binds to laminin and abrogates the effect of laminin on cell morphology and adhesion. Exp Cell Res 173:193–205

Bracke ME, Boterberg T, Bruyneel EA, Mareel MM (1999) Collagen invasion assay. In: Brooks S, Schumacher E (eds) Metastasis methods and protocols. Humana Press, Totowa (in press)

Bruyneel E, Mareel M (1981) Early activities of invasive malignant cells in vitro. Arch Geschwulstforsch 51:34–39

Buddecke E (1984) Grundriss der Biochemie. Walter de Gruyter, Berlin

Chan BMC, Matsuura N, Takada Y, Zetter BR, Hemler ME (1991) In vitro and in vivo consequences of VLA-2 expression on rhabdomyosarcoma cells. Science 251:1600–1602

Chen W, Öbrink B (1991) Cell-cell contacts mediated by E-cadherin (Uvomorulin) restrict invasive behavior of L-cells. J Cell Biol 114:319–327

Chen WY, Tseng SC (1995) Differential intrastromal invasion by normal ocular surface epithelia is mediated by different fibroblasts. Exp Eye Res 61:521–534

Chintala SK, Gokaslan ZL, Go Y, Sawaya R, Nicolson GL, Rao JS (1996) Role of extracellular matrix proteins in regulation of human glioma cell invasion in vitro. Clin Exp Metastasis 14:358–366

Crepaldi T, Gautreau A, Comoglio PM, Louvard D, Arpin M (1997) Ezrin is an effector of hepatocyte growth factor-mediated migration and morphogenesis in epithelial cells. J Cell Biol 138:423–434

Date K, Matsumoto K, Kuba K, Shimura H, Tanaka M, Nakamura T (1998) Inhibition of tumor growth and invasion by a four-kringle antagonist (HGF/NK4) for hepatocyte growth factor. Oncogene 17:3045–3054

Davis GE (1992) Affinity of integrins for damaged extracellular matrix: $\alpha_v\beta_3$ binds to denatured collagen type I through RGD sites. Biochem Biophys Res Commun 182:1025–1031

De Bruyne GK, Bracke ME, Plessers L, Mareel MM (1988) Invasiveness in vitro of mixed aggregates composed of two human mammary cell lines MCF-7 and HBL-100. Invasion Metastasis 8:253–265

Delannoy-Courdent A, Mattot V, Fafeur V, Fauquette W, Pollet I, Calmels T, Vercamer C, Boilly B, Vandenbunder B, Desbiens X (1998) The expression of an Ets1 transcription factor lacking its activation domain decreases uPA proteolytic activity and cell motility, and impairs normal tubulogenesis and cancerous scattering in mammary epithelial cells. J Cell Sci 111:1521–1534

Del Buono R, Pignatelli M, Bodmer WF, Wright NA (1991) The role of the arginine-glycine-aspartic acid-directed cellular binding to type I collagen and rat mesenchymal cells in colorectal tumour differentiation. Differentiation 46:97–103

de Ridder L, Bruyneel E, Calliauw L (1994) The invasiveness in vitro of brain-tumour derived cells depends on their micro-ecosystem. Acta Neurochir 130:140–143

Deroanne CF, Hajitou A, Calberg-Bacq CM, Nusgens BV, Lapière CM (1997) Angiogenesis by fibroblast growth factor 4 is mediated through an autocrine up-regulation of vascular endothelial growth factor expression. Cancer Res 57:5590–5597

Deryugina EI, Bourdon MA, Reisfeld RA, Strongin A (1998) Remodeling of collagen matrix by human tumor cells requires activation and cell surface association of matrix metalloproteinase-2. Cancer Res 58:3743–3750

Docherty RJ, Forrester JV, Lackie JM (1987) Type I collagen permits invasive behaviour by retinal pigmented epithelial cells in vitro. J Cell Sci 87:399–409

Docherty R, Forrester JV, Lackie JM, Gregory DW (1989) Glycosaminoglycans facilitate the movement of fibroblasts through three-dimensional collagen matrices. J Cell Sci 92:263–270

Doki Y, Shiozaki H, Tahara H, Inoue M, Oka H, Iihara K, Kadowaki T, Takeichi M, Mori T (1993) Correlation between E-cadherin expression and invasiveness in vitro in a human esophageal cancer cell line. Cancer Res 53:3421–3426

Ellerbroek SM, Fishman DA, Kearns AS, Bafetti LM, Stack MS (1999) Ovarian carcinoma regulation of matrix metalloproteinase-2 and membrane type 1 matrix metalloproteinase through $\beta 1$ integrin. Cancer Res 59:1635–1641

Empereur S, Djelloul S, Di Gioia Y, Bruyneel E, Mareel M, Van Hengel J, Van Roy F, Comoglio P, Courtneidge S, Paraskeva C, Chastre E, Gespach C (1997) Progression of familial adenomatous polyposis (FAP) colonic cells after transfer of the src or polyoma middle T oncogenes: cooperation between src and HGF/Met in invasion. Br J Cancer 75:241–250

Erkell LJ, Schirrmacher V (1988) Quantitative in vitro assay for tumor cell invasion through extracellular matrix or into protein gels. Cancer Res 48:6933–6937

Festuccia C, Bologna M, Gravina GL, Guerra F, Angelucci A, Villanova I, Millimaggi D, Teti A (1999) Osteoblast-conditioned media contain TGF-β1 and modulate the migration of prostate tumor cells and their interactions with extracellular matrix components. Int J Cancer 81:395–403

Fidler IJ (1990) Critical factors in the biology of human cancer metastasis: twenty-eighth G.H.A. Clowes award memorial award lecture. Cancer Res 50:6130–6138

Fisher C, Gilbertson-Beadling S, Powers EA, Petzold G, Poorman R, Mitchell MA (1994) Interstitial collagenase is required for angiogenesis in vitro. Dev Biol 162:499–510

Friedl P, Maaser K, Klein CE, Niggemann B, Krohne G, Zänker KS (1997) Migration of highly aggressive MV3 melanoma cells in 3-dimensional collagen lattices results in local matrix reorganization and shedding of α2 and β1 integrins and CD44. Cancer Res 57:2061–2070

Frixen UH, Nagamine Y (1993) Stimulation of urokinase-type plasminogen activator expression by blockage of E-cadherin-dependent cell-cell adhesion. Cancer Res 53:3618–3623

Frixen UH, Behrens J, Sachs M, Eberle G, Voss B, Warda A, Löchner D, Birchmeier W (1991) E-cadherin-mediated cell-cell adhesion prevents invasiveness of human carcinoma cells. J Cell Biol 113:173–185

Fujii K, Furukawa F, Matsuyoshi N (1996) Ligand activation of overexpressed epidermal growth factor receptor results in colony dissociation and disturbed E-cadherin function in HSC-1 human cutaneous squamous carcinoma cells. Exp Cell Res 223:50–62

Gaetje R, Kotzian S, Herrmann G, Baumann R, Starzinski-Powitz A (1995) Invasiveness of endometriotic cells in vitro. Lancet 346:1463–1464

Gaetje R, Kotzian S, Herrmann G, Baumann R, Starzinski-Powitz A (1997) Nonmalignant epithelial cells, potentially invasive in human endometriosis, lack the tumor suppressor molecule E-cadherin. Am J Pathol 150:461–467

Gajdusek CM, Luo Z, Mayberg MR (1993) Basic fibroblast growth factor and transforming growth factor beta-1: synergistic mediators of angiogenesis in vitro. J Cell Physiol 157:133–144

Galzie Z, Fernig DG, Smith JA, Poston GJ, Kinsella AR (1997) Invasion of human colorectal carcinoma cells is promoted by endogenous basic fibroblast growth factor. Int J Cancer 71:390–395

Gamble JR, Matthias LJ, Meyer G, Kaur P, Russ G, Faull R, Berndt MC, Vadas MA (1993) Regulation of in vitro capillary tube formation by anti-integrin antibodies. J Cell Biol 121:931–943

Gilles C, Bassuk JA, Pulyaeva H, Sage EH, Foidart JM, Thompson EW (1998) SPARC/osteonectin induces matrix metalloproteinase 2 activation in human breast cancer cell lines. Cancer Res 58:5529–5536

Goto F, Goto K, Weindel K, Folkman J (1993) Synergistic effects of vascular endothelial growth factor and basic fibroblast growth factor on the proliferation and cord formation of bovine capillary endothelial cells within collagen gels. Lab Invest 69:508–517

Grinnell F (1994) Mini-review on the cellular mechanisms of disease. Fibroblasts, myofibroblasts, and wound contraction. J Cell Biol 124:401–404

Gullberg D, Gehlsen KR, Turner DC, Åhlén K, Zijenah LS, Barnes MJ, Rubin K (1992) Analysis of $\alpha_1\beta_1$, $\alpha_2\beta_1$ and $\alpha_3\beta_1$ integrins in cell-collagen interactions: identification of conformation dependent $\alpha_1\beta_1$ binding sites in collagen type I. EMBO J 11:3865–3873

Hajitou A, Baramova EN, Bajou K, Noë V, Bruyneel E, Mareel M, Collette J, Foidart JM, Calberg-Bacq CM (1998) FGF-3 and FGF-4 elicit distinct oncogenic properties in mouse mammary myoepithelial cells. Oncogene 17:2059–2071

Hatta K, Takeichi M (1986) Expression of N-Cadherin adhesion molecules associated with early morphogenetic events in chick development. Nature 320:447–449

Hirai Y, Lochter A, Galosy S, Koshida S, Niwa S, Bissell MJ (1998) Epimorphin functions as a key morphoregulator for mammary epithelial cells. J Cell Biol 140:159–169

Ilan N, Mahooti S, Madri JA (1998) Distinct signal transduction pathways are utilized during the tube formation and survival phases of in vitro angiogenesis. J Cell Sci 111:3621–3631

Infeld MD, Brennan JA, Davis PB (1993) Human fetal lung fibroblasts promote invasion of extracellular matrix by normal human tracheobronchial epithelial cells in vitro: a model of early airway gland development. Am J Respir Cell Mol Biol 8:69–76

Iwazawa T, Shiozaki H, Doki Y, Inoue M, Tamura S, Matsui S, Monden T, Matsumoto K, Nakamura T, Monden M (1996) Primary human fibroblasts induce diverse tumor invasiveness: involvement of HGF as an important paracrine factor. Jpn J Cancer Res 87:1134–1142

Jackson CJ, Jenkins K, Schrieber L (1992) Possible mechanisms of type I collagen-induced vascular tube formation. In: Steiner R, Weisz PB, Langer R (eds) Angiogenesis: key principles–science–technology–medicine. Birkhäuser, Basel, pp 198–204

Kadler KE, Holmes DF, Trotter JA, Chapman JA (1996) Collagen fibril formation. Biochem J 316:1–11

Kanda S, Tomasini-Johansson B, Klint P, Dixelius J, Rubin K, Claesson-Welsh L (1999) Signaling via fibroblast growth factor receptor-1 is dependent on extracellular matrix in capillary endothelial cell differentiation. Exp Cell Res 248:203–213

Katz SG (1995) Extracellular and intracellular degradation of collagen by trophoblast giant cells in acute fasted mice examined by electron microscopy. Tissue Cell 27:713–721

Kawahara E, Okada Y, Nakanishi I, Iwata K, Kojima S, Kumagai S, Yamamoto E (1993) The expression of invasive behavior of differentiated squamous carcinoma cell line evaluated by an in vitro invasion model. Jpn J Cancer Res 84(4):409–418

Keirsebilck A, Bonné S, Bruyneel E, Vermassen P, Lukanidin E, Mareel M, van Roy F (1998) E-cadherin and metastasin (mts-1/S100A4) expression levels are inversely regulated in two tumor cell families. Cancer Res 58:4587–4591

Kelly T, Yan Y, Osborne RL, Athota AB, Rozypal TL, Colclasure JC, Chu WS (1998) Proteolysis of extracellular matrix by invadopodia facilitates human breast cancer cell invasion and is mediated by matrix metalloproteinases. Clin Exp Metastasis 16:501–512

Kim K, Daniels KJ, Hay ED (1998) Tissue-specific expression of β-catenin in normal mesenchyme and uveal melanomas and its effect on invasiveness. Exp Cell Res 245:79–90

Kinsella AR, Lepts GC, Hill CL, Jones M (1994) Reduced E-cadherin expression correlates with increased invasiveness in colorectal carcinoma cell lines. Clin Exp Metastasis 12:335–342

Klein CE, Dressel D, Steinmayer T, Mauch C, Eckes B, Krieg T, Bankert RB, Weber L (1991) Integrin $\alpha 2\beta 1$ is upregulated in fibroblasts and highly aggressive melanoma cells in three-dimensional collagen lattices and mediates the reorganization of collagen I fibrils. J Cell Biol 115:1427–1436

Kleinman HK, McGarvey ML, Hassell JR, Star VL, Cannon FB, Laurie GW, Martin GR (1986) Basement membrane complexes with biological activity. Biochemistry 25:312–318

Knäuper V, Cowell S, Smith B, López-Otin C, O'Shea M, Morris H, Zardi L, Murphy G (1997) The role of the C-terminal domain of human collagenase-3 (MMP-13) in the activation of procollagenase-3, substrate specificity, and tissue inhibitor of metalloproteinase interaction. J Biol Chem 272:7608–7616

Komada N, Nabeshima K, Koita H, Kataoka H, Muraoka K, Koone M (1993) Characteristics of a metastatic variant to the liver of human rectal adenocarcinoma cell line RCM-1. Invasion Metastasis 13:38–49

Kono T, Furukawa M, Tanii T, Taniguchi S, Mizuno N, Ishii M, Hamada T (1990) Infiltration of melanoma cells into the type I collagen gel. J Dermatol 17:473–476

Kotelevets L, Noë V, Bruyneel E, Myssiakine E, Chastre E, Mareel M, Gespach C (1998) Inhibition by platelet-activating factor of Src- and hepatocyte growth factor-dependent invasiveness of intestinal and kidney epithelial cells. Phosphatidylinositol-3′-kinase is a critical mediator of tumor invasion. J Biol Chem 273:14138–14145

Koutsilieris M, Sourla A, Pelletier G, Doillon CJ (1994) Three-dimensional type I collagen gel system for the study of osteoblastic metastases produced by metastatic prostate cancer. J Bone Mineral Res 9:1823–1832

Kubota Y, Kleinman HK, Martin GR, Lawley TJ (1988) Role of laminin and basement membrane in the morphological differentiation of human endothelial cells into capillary-like structures. J Cell Biol 107:1589–1598

Kuzuya M, Satake S, Ai S, Asai T, Kanda S, Ramos MA, Miura H, Ueda M, Iguchi A (1998) Inhibition of angiogenesis on glycated collagen lattices. Diabetologia 41:491–499

Langholz O, Röckel D, Mauch C, Kozlowska E, Bank I, Krieg T, Eckes B (1995) Collagen and collagenase gene expression in three-dimensional collagen lattices are differentially regulated by $\alpha 1\beta 1$ and $\alpha 2\beta 1$ integrins. J Cell Biol 131:1903–1915

Langholz O, Roeckel D, Petersohn D, Broermann E, Eckes B, Krieg T (1997) Cell-matrix interactions induce tyrosine phosporylation of MAP kinases ERK1 and ERK2 and PLCγ-1 in two-dimensional and three-dimensional cultures of human fibroblasts. Exp Cell Res 235:22–27

Laoide BM, Gastinne I, Rougeon F (1999) Tubular morphogenesis and mesenchymal interactions affect renin expression and secretion in SIMS mouse submandibular cells. Exp Cell Res 248:172–185

Leavesley DI, Schwartz MA, Rosenfeld M, Cheresh DA (1993) Integrin $\beta 1$- and $\beta 3$-mediated endothelial cell migration is triggered through distinct signaling mechanisms. J Cell Biol 121:163–170

Le Marer N, Bruyneel E (1996) Comparison of in vitro invasiveness of human breast carcinoma in early or late stage with their malignancy in vivo. Anticancer Res 16:2767–2772

Leroy A, Mareel M, Nelis H (1997) Metastasis of bacteria, parasites, leukocytes and cancer cells. In: Bryne M, Nestland JM (eds) Pathology update, vol 4: The metastatic process. Gustav Fisher, Jena, pp 21–40

Li H, Shimura H, Aoki Y, Date K, Matsumoto K, Nakamura T, Tanaka M (1998) Hepatocyte growth factor stimulates the invasion of gallbladder carcinoma cell lines in vitro. Clin Exp Metastasis 16:74–82

Linsenmayer (1981) Collagen. In: Hay ED (ed) Cell biology of extracellular matrix. Plenum Press, New York, pp 5–37

Liotta LA, Rao CN, Barsky SH (1983) Tumor invasion and the extracellular matrix. Lab Invest 49:636–649

Liotta LA, Steeg PS, Stetler-Stevenson WG (1991) Cancer metastasis and angiogenesis: an imbalance of positive and negative regulation. Cell 64:327–336

Liu W, Litwack ED, Stanley MJ, Langford JK, Lander AD, Sanderson RD (1998) Heparan sulfate proteoglycans as adhesive and anti-invasive molecules. Syndecans and glypican have distinct functions. J Biol Chem 273:22825–22832

Lochter A, Srebrow A, Sympson CJ, Terracio N, Werb Z, Bissell MJ (1997) Misregulation of stromelysin-1 expression in mouse mammary tumor cells accompanies acquisition of stromelysin-1-dependent invasive properties. J Biol Chem 272:5007–5015

Mareel MM, De Baetselier P, Van Roy FM (1991) Mechanisms of invasion and metastasis. CRC Press, Boca Raton

Mareel MM, Van Roy FM, Bracke ME (1993) How and when do tumor cells metastasize? Crit Rev Oncog 4:559–594

Mareel M, Van Roy F, De Baetselier P, Vakaet L (1994) Invasiveness in development and neoplasia. In: Hodges GM, Rowlatt C (eds) Developmental biology and cancer. Telford Press, New Jersey, pp 389–416

Martel C, Harper F, Cereghini S, Noë V, Mareel M, Crémisi C (1997) Inactivation of retinoblastoma family proteins by SV40 T antigen results in creation of a hepatocyte growth factor/scatter factor autocrine loop associated with an epithelial-fibroblastoid conversion and invasiveness. Cell Growth Differ 8:165–178

Matsumoto K, Horikoshi M, Rikimaru K, Enomoto S (1989) A study of an in vitro model for invasion of oral squamous cell carcinoma. J Oral Pathol Med 18:498–501

Matsuyoshi N, Hamaguchi M, Taniguchi S, Nagafuchi A, Tsukita S, Takeichi M (1992) Cadherin-mediated cell-cell adhesion is perturbed by v-*src* tyrosine phosphorylation in metastatic fibroblasts. J Cell Biol 118:703–714

McCarthy JB, Vachhani B, Iida J (1996) Cell adhesion to collagenous matrices. Biopolymers 40:371–381

Meyer MB, Bastholm L, Nielsen MH, Elling F, Rygaard J, Chen W, Öbrink B, Bock E, Edvardsen K (1995) Localization of NCAM on NCAM-B-expressing cells with inhibited migration in collagen. APMIS 103:197–208

Montesano R, Orci L (1985) Tumor-promoting phorbol esters induce angiogenesis in vitro. Cell 42:469–477

Montesano R, Vassalli JD, Baird A, Guillemin R, Orci L (1986) Basic fibroblast growth factor induces angiogenesis in vitro. Proc Natl Acad Sci USA 83:7297–7301

Montesano R, Matsumoto K, Nakamura T, Orci L (1991) Identification of a fibroblast-derived epithelial morphogen as hepatocyte growth factor. Cell 67:901–908

Montesano R, Pepper MS, Orci L (1993) Paracrine induction of angiogenesis in vitro by Swiss 3T3 fibroblasts. J Cell Sci 105:1013–1024

Montesano R, Kumar S, Orci L, Pepper MS (1996) Synergistic effect of hyaluronan oligosaccharides and vascular endothelial growth factor on angiogenesis in vitro. Lab Invest 75:249–262

Mooradian DL, McCarthy JB, Komanduri KV, Furcht LT (1992) Effects of transforming growth factor-β1 on human pulmonary adenocarcinoma cell adhesion, motility, and invasion in vitro. J Natl Cancer Inst 84:523–527

Muir D (1995) Differences in proliferation and invasion by normal, transformed and NF1 Schwann cell cultures are influenced by matrix metalloproteinase expression. Clin Exp Metastasis 13:303–314

Murray RK, Keeley FW (1993) The extracellular matrix. In: Murray RK, Granner DK, Mayes PA, Rodwell VW (eds) Harper's biochemistry. Appleton & Lange, East Norwalk, Connecticut, pp 634–646

Nabeshima K, Kataoka H, Koono M (1986) Enhanced migration of tumor cells in response to collagen degradation products and tumor cell collagenolytic activity. Invasion Metastasis 6:270–286

Nakayama Y, Naito S, Ryuto M, Hata Y, Ono M, Sueishi K, Komiyama S, Itoh H, Kuwano M (1996) An in vitro invasion model for human renal cell carcinoma cell lines mimicking their metastatic abilities. Clin Exp Metastasis 14:466–474

Nathan C, Sporn M (1991) Mini-review: cytokines in context. J Cell Biol 113:981–986

Noë V, Willems J, Vandekerckhove J, Van Roy F, Bruyneel E, Mareel M (1999) Inhibition of adhesion and induction of epithelial cell invasion by HAV-containing E-cadherin-specific peptides. J Cell Sci 112:127–135

Oft M, Peli J, Rudaz C, Schwarz H, Beug H, Reichmann E (1996) TGF-β1 and Ha-Ras collaborate in modulating the phenotypic plasticity and invasiveness of epithelial tumor cells. Genes Dev 10:2462–2477

Oft M, Heider KH, Beug H (1998) TGFβ signaling is necessary for carcinoma cell invasiveness and metastasis. Curr Biol 8:1243–1252

Ono M, Okamura K, Nakayama Y, Tomita M, Sato Y, Komatsu Y, Kuwano M (1992) Induction of human microvascular endothelial tubular morphogenesis by human keratinocytes: involvement of transforming growth factor-α. Biochem Biophys Res Commun 189:601–609

Prunier C, Mazars A, Noë V, Bruyneel E, Mareel M, Gespach C, Atfi A (1999) Evidence that Smad2 is a tumour suppressor implicated in the control of cellular invasion. J Biol Chem 274:22919–22922

Ravanti L, Heino J, López-Otín C, Kähäri VM (1999) Induction of collagenase-3 (MMP-13) expression in human skin fibroblasts by three-dimensional collagen is mediated by p38 mitogen-activated protein kinase. J Biol Chem 274:2446–2455

Roeckel D, Krieg T (1994) Three-dimensional contact with type I collagen mediates tyrosine phosphorylation in primary human fibroblasts. Exp Cell Res 211:42–48

Rosenthal EL, Johnson TM, Allen ED, Apel IJ, Punturieri A, Weiss SJ (1998) Role of the plasminogen activator and matrix metalloproteinase systems in epidermal growth factor- and scatter factor-stimulated invasion of carcinoma cells. Cancer Res 58:5221–5230

Sakakura T (1991) New aspects of stroma-parenchyma relations in mammary gland differentiation. Int Rev Cytol 125:165–202

Sander EE, van Delft S, ten Klooster JP, Reid T, van der Kammen RA, Michiels F, Collard JG (1998) Matrix-dependent Tiam1/Rac signaling in epithelial cells promotes either cell-cell adhesion or cell migration and is regulated by phosphatidylinositol 3-kinase. J Cell Biol 143:1385-1398

Sankar S, Mahooti-Brooks N, Bensen L, McCarthy TL, Centrella M, Madri JA (1996) Modulation of transforming growth factor β receptor levels on microvascular endothelial cells during in vitro angiogenesis. J Clin Invest 97:1436-1446

Sato H, Takino T, Okada Y, Cao J, Shinagawa A, Yamamoto E, Seiki M (1994) A matrix metallo-proteinase expressed on the surface of invasive tumour cells. Nature 370:61-65

Sato T, Foged NT, Delaisse JM (1998) The migration of purified osteoclasts through collagen is inhibited by matrix metalloproteinase inhibitors. J Bone Miner Res 13:59-66

Schaller MD, Otey CA, Hildebrand JD, Parsons JT (1995) Focal adhesion kinase and paxillin bind to peptides mimicking β integrin cytoplasmic domains. J Cell Biol 130:1181-1187

Schirren CG, Scharffetter K, Hein R, Braun-Falco O, Krieg T (1990) Tumor necrosis factor α induces invasiveness of human skin fibroblasts in vitro. Invest Dermatol 94:706-710

Schlaepfer DD, Hanks SK, Hunter T, van der Geer P (1994) Integrin-mediated signal transduction linked to Ras pathway by GRB2 binding to focal adhesion kinase. Nature 372:786-790

Schönherr E, O'Connell BC, Schittny J, Robenek H, Fastermann D, Fisher LW, Plenz G, Vischer P, Young MF, Kress H (1999) Paracrine or virus-mediated induction of decorin expression by endothelial cells contributes to tube formation and prevention of apoptosis in collagen lattices. Eur J Cell Biol 78:44-55

Schor SL (1980) Cell proliferation and migration on collagen substrata in vitro. J Cell Sci 41:159-175

Schor SL, Schor AM (1987) Foetal-to-adult transitions in fibroblast phenotype: their possible relevance to the pathogenesis of cancer. J Cell Sci Suppl 8:165-180

Schor SL, Schor AM, Winn B, Rushton G (1982) The use of three-dimensional collagen gels for the study of tumor cell invasion in vitro: experimental parameters influencing cell migration into the gel matrix. Int J Cancer 29:57-62

Schor SL, Schor AM, Allen TD, Winn B (1985a) The interaction of melanoma cells with fibroblasts and endothelial cells in three-dimensional macromolecular matrices: a model for tumour cell invasion. Int J Cancer 36:93-102

Schor SL, Schor AM, Rushton G, Smith L (1985b) Adult, foetal and transformed fibroblasts display different migratory phenotypes on collagen gels: evidence for an isoformic transition during foetal development. J Cell Sci 73:221-234

Schor SL, Ellis I, Dolman C, Banyard J, Humphries MJ, Mosher DF, Grey AM, Mould AP, Sottile J, Schor AM (1996) Substratum-dependent stimulation of fibroblast migration by the gelatin-binding domain of fibronectin. J Cell Sci 109:2581-2590

Seltzer JL, Lee AY, Akers KT, Sudbeck B, Souton EA, Wayner EA, Eisen AZ (1994) Activation of 72-kDa type IV collagenase/gelatinase by normal fibroblasts in collagen lattices is mediated by integrin receptors but is not related to lattice contraction. Exp Cell Res 213:365-374

Shimura H, Date K, Matsumoto K, Nakamura T, Tanaka M (1995) Induction of invasive growth in a gallbladder cancer cell line by hepatocyte growth factor in vitro. Jpn J Cancer Res 86:662-669

Skobe M, Rockwell P, Goldstein N, Vosseler S, Fusenig NE (1997) Halting angiogenesis suppresses carcinoma cell invasion. Nat Med 3:1222-1227

Soriano JV, Pepper MS, Nakamura T, Orci L, Montesano R (1995) Hepatocyte growth factor stimulates extensive development of branching duct-like structures by cloned mammary gland epithelial cells. J Cell Sci 108:413-430

Soriano JV, Orci L, Montesano R (1996) TGF-β1 induces morphogenesis of branching cords by cloned mammary epithelial cells at subpicomolar concentrations. Biochem Biophys Res Commun 220:879-885

Sourla A, Doillon C, Koutsilieris M (1996) Three-dimensional type I collagen gel system containing MG-63 osteoblasts-like cells as a model for studying local bone reaction caused by metastatic cancer cells. Anticancer Res 16:2773-2780

Spieker N, Mareel M, Bruyneel EA, Nederbragt H (1995) E-cadherin expression and in vitro invasion of canine mammary tumor cells. Eur J Cell Biol 68:427–436

Stadler E, Dziadek M (1996) Extracellular matrix penetration by epithelial cells is influenced by quantitative changes in basement membrane components and growth factors. Exp Cell Res 229:360–369

Strohmaier AR, Spring H, Spiess E (1996) Three-dimensional analysis of the substrate-dependent invasive behavior of a human lung tumor cell line with a confocal laser scanning microscope. Histochem Cell Biol 105:179–185

Sweeney SM, Guy CA, Fields GB, San Antonio JD (1998) Defining the domains of type I collagen involved in heparin-binding and endothelial tube formation. Proc Natl Acad Sci USA 95:7275–7280

Takeichi M (1995) Morphogenetic roles of classic cadherins. Curr Opin Cell Biol 7:619–627

Tamaki M, McDonald W, Amberger VR, Moore E, Del Maestro RF (1997) Implantation of C6 astrocytoma spheroid into collagen type I gels: invasive, proliferative, and enzymatic characterizations. J Neurosurg 87:602–609

Tucker GC, Boyer B, Gavrilovic J, Emonard H, Thiery JP (1990) Collagen-mediated dispersion of NBT-II rat bladder carcinoma cells. Cancer Res 50:129–137

Turksen K, Choi Y, Fuchs E (1991) Transforming growth factor alpha induces collagen degradation and cell migration in differentiating human epidermal raft cultures. Cell Regul 2:613–625

Uyttendaele H, Soriano JV, Montesano R, Kitajewski J (1998) Notch4 and Wnt-1 proteins function to regulate branching morphogenesis of mammary epithelial cells in an opposing fashion. Dev Biol 196:204–217

Vakaet L Jr, Vleminckx K, Van Roy F, Mareel M (1991) Numerical evaluation of the invasion of closely related cell lines into collagen type I gels. Invasion Metastasis 11:249–260

Vernon RB, Sage EH (1999) A novel, quantitative model for study of endothelial cell migration and sprout formation within three-dimensional collagen matrices. Microvasc Res 57:118–133

Vleminckx K, Vakaet L Jr, Mareel M, Fiers W, Van Roy F (1991) Genetic manipulation of E-cadherin expression by epithelial tumor cells reveals an invasion suppressor role. Cell 66:107–119

Wach F, Eyrich AM, Wustrow T, Krieg T, Hein R (1996) Comparison of migration and invasiveness of epithelial tumor and melanoma cells in vitro. J Dermatol Sci 12:118–126

Wang TN, Qian XH, Granick MS, Solomon MP, Rothman VL, Tuszynski GP (1995) The effect of thrombospondin on oral squamous carcinoma cell invasion of collagen. Am J Surg 170:502–505

Wang TN, Qian X, Granick MS, Solomon MP, Rothman VL, Berger DH, Tuszynski GP (1996) Thrombospondin-1 (TSP-1) promotes the invasive properties of human breast cancer. J Surg Res 63:39–43

Wordinger RJ, Brun-Zinkernagel AM, Jackson T (1991) An ultrastructural study of in-vitro interaction of guinea-pig and mouse blastocysts with extracellular matrices. J Reprod Fertil 93:585–587

Xu J, Clark RAF (1997) A three-dimensional collagen lattice induces protein kinase C-ζ activity: role in α_2 integrin and collagenase mRNA expression. J Cell Biol 136:473–483

Yamada S, Toda S, Shin T, Sugihara H (1999) Effects of stromal fibroblasts and fat cells and an environmental factor air exposure on invasion of laryngeal carcinoma (HEp-2) cells in a collagen gel invasion assay system. Arch Otolaryngol Head Neck Surg 125:424–431

Zamora PO, Danielson KG, Hosick HL (1980) Invasion of endothelial cell monolayers on collagen gels by cells from mammary tumor spheroids. Cancer Res 40:4631–4639

Subject Index

FSC
www.fsc.org
MIX
Papier aus verantwortungsvollen Quellen
Paper from responsible sources
FSC® C105338